All my best
Jake 2020

Gary Dumond Remembers

Maine Warden Pilots

In CONVERSATION with Jake Morrel

ISBN: 978-1-63381-145-4

All the photos are from the Malcolm Maheu collection

Designed and Produced by
Maine Authors Publishing
12 High Street, Thomaston, Maine 04861
www.maineauthorspublishing.com

Printed in the United States of America

INTRODUCTION

World War II set the stage for a surge in bush flying when the conflict ended. The airplane manufacturers had produced thousands of small observation planes for the military which had trained thousands of pilots. In northern Maine numerous flying services opened and used small float planes to transport fishermen, hunters, woodsmen, and freight into the wilderness.

The Maine Warden Service also bought seaplanes and hired several new warden pilots, among them Malcolm "Mac" Maheu, George Later, Wilfred "Sleepy" Atkins, and Andy Stinson. Together they patrolled the Maine skies for over twenty years before Gary Dumond was hired in 1972. Gary flew more than twenty thousand hours before he retired in 1991.

Gary's career overlapped his mentors by several years, and he heard and reheard their stories of adventures in the '50s and '60s.

On these pages those tales of aerial searches, airplane mishaps, wildlife tracking, and much more come to life. Gary also tells many of his own and explains in detail the life of a Maine Warden Pilot.

In the winter of 2018 Gary and I had a lengthy recorded conversation where he shared those memories. What follows is a transcription of that dialog.

—Jake Morrel

Gary Dumond

EAGLE LAKE, MAINE, MARCH 19, 2018

So, we're sitting here in Eagle Lake, Maine, ready to have a conversation about Gary Dumond's flying career as a pilot for the Maine Warden Service. We're going to talk a lot about the people that came before Gary, some of the old planes, and some of the incidents that occurred. What I thought we might do first—I'm curious how you first got your flying bug. That is, when did you finally reach the point where you knew you wanted to make flying a career?

Well, Jake, I don't remember not being around airplanes. My dad had a flying service. I was born in 1950. He was running a flying service then. I can't remember not having an airplane for the family. So I was an airplane nut, and I'm still an airplane nut. When I studied at high school, I knew I wanted to fly and I didn't think I wanted to go to college. I put in to go to army flight school. They had a program called "high school to flight school." I lucked out and qualified, and to make a short story of it, I ended up going to army flight school to learn to fly helicopters. Went to Vietnam in 1970; all of 1970, I was in Vietnam. I flew fourteen hundred and ten hours in eleven and one half months. Got back from Vietnam, got married, went to work for Aroostook Airways in Presque Isle for a few months in 1971. In the summer of '71, Maine Forestry had a pilot, Glen Sherman. He had to get done flying for a medical reason. They needed a pilot and I applied for the job and I got the job as a pilot in Old Town. So I worked there for about a year and a half. I flew both the Beaver and the Cessnas, plus a lot of the helicopter flying. And along in the fall of '72, the Maine Warden Service lost a pilot in an accident, and needed a pilot. It was close to hunting season, and they needed a guy that fall, right away. So they called me up and wanted to know if I still wanted to be a warden pilot. 'Course, that didn't take long! And I went to work in the fall of '72 as a game warden pilot.

That's interesting! One thing—we're really lucky that happened [because] you overlapped with the oldest of the Maine Fish and Game pilots, the guys who started right after World War II. And I'm sure they were important in what you know now. Let's talk about some of the old-time pilots, the ones you started with, and what you can remember about them and the airplanes they flew.

Well, I can tell you that I knew who those warden pilots were. I met Mac Maheu when I was probably eight or nine years old. He was the pilot here in Eagle Lake, and then Dick Varney came. My dad used to come over here to Eagle Lake. He was a Maine State Police officer, and they'd have a get-together, a bunch of the Troopers and wardens in the camp where you are staying. And I'd go out and sit in that 180 in the middle of the winter. That was before the hangar. The airplane was tied out and I'd sit in that airplane for an hour at a time. That was out of this world—you know! So all the early pilots, even if I had never met them, I'd heard stories. I knew who Andy Stinson was. 'Course, I knew Mac because I'd met him a few times. And I knew who George Later was. The only early pilots that I never met were Bill Turgeon, the first pilot, and George Townsend, who was one of the first pilots and worked out of Rangeley. But other than that, I ended up working with those guys. When I went to work, George Later was still the chief pilot. I worked with Andy and I worked with Mac, who was just retiring. I worked with Dick Varney. Matter of fact, I helped Dick

Varney, Andy, and George learn to fly helicopters when I was still working for the Maine Forestry.

So when those guys started in 1946, or whenever Mac started, what did they give them for airplanes?

The original plane they had was a Gullwing Stinson that was given to them by one of the commissioners. And of course, Bill Turgeon flew that airplane. Then they had Piper PA-12s. There was a big to-do there, and I think at one time they maybe had four of them. That was the original ones with the 108-horsepower engine, and with a little modification, they got 115 horse out of them. They didn't have flaps [for slower flight] or anything like that, but it was a pretty good airplane for what they were doing. Then they got into Aeronca Sedans. So, a combination of Aeroncas and PA-12s. The PA-12s, after a while they upped to 125 horse and they put flaps on them. That became a pretty good airplane. That went along for a while, and then they got their first Cessna 180 in about '55. Mac had the first 180.

In the very beginning, did George or Mac have any Piper J-3 Cubs?

George was a district game warden assigned to Ripogenus Dam. And I think they leased a Piper J-3 and George flew that one winter before he was actually a warden pilot. He flew that on skis. I don't know that the department ever owned a J-3 airplane. I think they may have leased a few or something. But the only time they had a J-3 that I know of was when George flew it.

You mentioned that they added flaps to the Cruiser [PA-12]. That's a big deal, both paperwork-wise and design-wise. Who was the person who was doing the paperwork and the engineering on it? Or did they just do it by trial and error?

They had to get everything certified, even back then. I know Howard Lambertson would work for them. He was working in Waterville for Marden Airways. Howard, of course, he became our mechanic down the road, the best of the best! They had some plans. The first Cruiser they did, Howard actually owned the STC [FAA design approval] on the flap installation. And that worked out pretty good. They ended up doing that to all the Cruisers.

Was the flap similar to a Super Cub flap? It wasn't a Fowler flap [more modern design] or anything like that?

No, it was similar to a Super Cub flap. When they started doing that, I want to say it would be in the early '50s…I don't know what airplane they got the flap from, but I've seen pictures of it. As a matter of fact, I saw a magazine one time, and I may still have it, and it had a story of PA-12s. And they had a picture and a diagram of a PA-12 and it showed a flap installation, and at the bottom it said, "This airplane was owned by Maine Fish and Game Department 1952" or whatever. And Howard Lambertson owned the STC. After that, there were quite a few STCs that came out for [installing flaps on] Super Cruisers.

In the beginning, say '40s or '50s, 'course they were flying planes much slower than the Cessna 180s or 185s of today. Did they have to stay a lot at outpost camps because they didn't have time to return before dark or whatever?

Oh, yeah; you know it was a completely different operation than when I was flying, and it was certainly different than it is now. I mean, all the different camps, a lot of them are gone now; some of them are still around. You know, they'd use them, they'd end up staying there because of weather. I can remember Mac telling me of flying into Clayton Lake with Curtis Cooper a warden supervisor.

They tied the airplane down, put the wing covers on, and they snowshoed to Daaquam. That's like fourteen miles. They went halfway, spent the night in a camp on the St. John River. The next day, they snowshoed the rest of the way into Daaquam. And of course, when they decided to come out, it was the same thing in reverse. I said to Mac, "Why didn't you land over there on the river somewhere?" He explained that they had Canadians doing illegal trapping on the American side, and the only way we were going to surprise them was to walk in. Imagine snowshoeing from Clayton Lake to Daaquam! And they knew where all the camps were. For a lot of the camps, everything that was in the camp was brought there by airplane.

And was it common that they would stash gas at those remote locations so they could operate out of there?

Oh, yeah, they did that a lot, and listen, the years that we flew the Super Cubs, Jack [McPhee] and I; that would have been '78 to '83 or four. We had to stash gas in different places. I mean, you could find gas here and there, Clayton Lake, or go into Munsungan or Shin Pond. But you needed gas, and when I was flying out of Old Town for those years, in Washington County there was no gas to be had. So every day when I left, I had four six-gallon cans in the back. When I got where I was going, to pick up a warden or whatever, I would leave cans there, and sometimes during the day I'd add a couple jugs, and maybe later, to get home, I'd add the last two. But in the early days, there was no gas around. There was no gas in Clayton Lake like there is today, so they had to have gas stored everywhere.

Now the other thing is they were operating in very cold weather conditions and spending the night. And I know from doing that myself how hard it is to get an airplane started in the morning. So later on we had what we called "red dragon heaters" that operated off twelve volts and needed a small propane cylinder. But I know in the early days they didn't have anything like that, so what was it like trying to get a Cub going at twenty below?

Andy Stinson, who was still stationed at Enfield, all those years never had the luxury of a hangar. He had the airplane tied down on Coldstream Pond, and he heated with a plumber's pot [torch]. And I remember watching him. I'd land there, going to meet him, and he'd have a big canvas hanging down around the nose and a plumber's pot underneath the airplane. And when he first got that started, there was flames everywhere. And when he warmed it up, he'd get underneath the canvas. He'd sit on a five-gallon pail, and he had to be careful or he'd get gassed out of there. And that's how they heated that for twenty-some years.

It's just amazing when you think of it: you've seen gas drip out of a gascolator [fuel strainer], and here you've got a flame right next to it. Did they ever set one on fire?

Not that I know of, but I mean, there have been airplanes set on fire. I know of one in Fort Kent when I was a kid. It was a Piper J-5, and the guy was heating it up and he turned around. When he turned back, the airplane was on fire and it burned so fast—'course it was nitrate dope, that all of a sudden the frame was sitting right there with no fabric on it.

You know, I started in Taylorcrafts. They were all hand propped. And we'd preheat them using similar methods. But one thing that happens a lot is, if they don't start, you can get a carburetor backfire and you have a little flame going. It seems weird, but what you have to do is keep cranking on the prop with the fire right next to you and suck it into the engine.

I saw a guy crank a 180, and that is just what happened. It backfired and flamed, and lucky he only burned some wires. He was up on Spider Lake but was able to fly back to Old Town.

Another thing that I actually have never done, but I know people who have done it, is to actually drain the oil out of the engine and then take it into the camp to keep it warm overnight. Then you dump it back in. Do you recall any of that?

I remember Jack doing that occasionally. He had a personal camp on Spider Lake. When he had the Cub, he told me he used to do that. When they went into the 185s, they ordered those with oil dilution from the gas. That was a touchy situation there, because when you used it for the first time, you'd come in from flying for the day then you'd dilute the oil. Then you had to drain it because it would wash a lot of the sludge and stuff out of the engine. Then you'd add oil and you were good for the rest of that oil change. Depending on how cold it was, that determined how long you'd dilute it. And what it did, it thinned the oil pretty good. That really wasn't the best of things, but that's all they had, you know. Nobody does that anymore. Now, they've all got the best of different types of heaters. This airplane I flew last winter, you could plug it in like a car. You could leave it plugged in all the time. If you don't fly for three days, when you start that airplane at twenty below, it's like starting it in June.

Another thing I'm curious about: Dick Folsom was my mentor, and he told me one thing right after another, whether it was about engines or how to fly the plane. With one sentence now and then, he probably saved my life. I'm really curious if the early guys like George talked to you about winter flying, how never to stop during taxi or whatever it was. Do you remember any of the things they told you that made a difference later on?

A lot of things. I was really lucky, you know. When I came to work, Mac was still around, George was chief pilot, and Andy was around, and Jack had been around then for a few years. If you were smart, you listened because there was a lot of good information. I can remember one of them saying, "You know, when you get up some morning and you figure you've got this airplane business all figured out, maybe you ought to take the day off—stay home that day because you're never going to have it all figured out." And I can remember Mac telling me, "December on the ice when you land somewhere, if you're not sure, keep the airplane going until you're right along shore where you know there's going to be good ice. And if you don't think there's any chance of it, unless you have to, you just don't land." But if you're moving along at six or seven miles an hour, an inch of ice will hold you up. When you stop, it becomes an issue.

Now, you know today quite often, since we have more roads, if you want to land someplace on skis, you can probably find somebody that's around to check the ice. In the beginning, it wasn't that way at all. When the first few trips were made in the winter, there were a lot of pilots landing on new ice for the first time since no one had given them any information. So what did you use as a guide early on when you would finally do it?

You know, we were lucky, because when I came to work there were radios, and if you were going to go and pick a guy up on Beau Lake up Estcourt way, you could ask him to check the ice along shore. 'Course, once you left your pickup truck, you couldn't talk to him anymore. That was before they had the portable radios. And I'd tell him, "You know, when you get there, check the ice, and if you don't think I should land, just cut a couple of big branches and make an X. You know, I'll see you standing there next to an X and leave and head back. And if there's nothing there, I'll assume it's good near shore." Later years, when we got the portable radios, you know, the wardens were really good, they knew what you had to have. They go out and they wouldn't make one hole, they'd probably make four or five holes. So you'd come in to land and you wouldn't stop until you got to where they were.

Did you ever ask them to put out spruce boughs for you so you could see the ice?

We used to do that, and especially on an overcast day, you can't see; it's so hard. It's kind of like snow blindness. It's kind of like glassy water. They would mark it with fir branches. I mean, we had guys, wardens, that lived in the back-country. I remember Portage Pond had a warden there. He'd have a runway that he maintained. He'd pack it down with a snow sled and he'd place some fir boughs. And in places we'd go into quite often, like Clayton Lake, they'd keep a runway marked. The problem with some places, as winter went on, was slush. When you fly, you can't look down and know there's slush—there's no way I know of telling that. And boy, slush, that is bad business! And I've been into it and I've been into it bad. I remember Mac telling me that he was stuck in the slush overnight on Musquacook; he had to spend the night. He got everything out, let it freeze down, had to take the skis off. They used to carry a jack with them in the airplane. They'd jack up one side and take the ski off, knock all the slush and ice off it, and put it back on. Then drop that one on fir boughs and jack up the other side. Can you imagine the amount of work they did back then?

Did you ever put engine oil on the bottom of your skis?

I never put on engine oil. You know, I was lucky, because by the time I came we started having the plastic bottoms. And what a difference! 'Course you get into slush, it don't make any difference what you have, you're going to freeze up. It's not just the bottom; the whole thing becomes a glob of ice. I got stuck in the slush up to Rangeley one year with…I can't remember who I had with me. But we were five hours getting out of there. Mistake I made there, I taxied toward shore where I was to meet this guy. And as I got closer to shore, I couldn't turn. And if I could have kept full power on, I'd have gone. I knew then I was in the slush and couldn't turn. Full power in a 185 you're going to get out of there. I learned something right there: you don't taxi right toward shore. You come parallel or at an angle.

You know, some people may not realize that slush can be a foot or more deep.

Oh, I know. They used to call Ross Lake the slush capital of the world. You know, in my days, in the early '70s, there'd be a lot of people out there ice fishing. And every year, we'd get calls, couple different times, "So-and-so didn't come home." And you'd go there, and there's the airplane stuck in the slush. They'd broken into someone's camp and they were there. Something about Ross Lake that caused a lot of slush. Back then, there seemed to be more snow, although there's plenty as we're talking right now. There were some years where slush was a real problem.

The other thing that some people may not realize, 'cause of the dams in Maine and the river systems, that even in our lakes in a lot of places called narrows, there's current. That can mean that a lake with two feet of ice on it will have open spots. What do you remember about some of the problems with narrows?

They're still a problem; that hasn't changed. For example, John's Bridge, between Big Eagle and Churchill…that narrows right there, I mean, sometimes people sled right under the bridge and right through. I've seen that many times, and you come back a week later and there was enough open water to land a floatplane! That's how fast it can change.

Yeah, I remember personally finding a deer struggling on the glare ice in the narrows to Lily Bay, and I decided I'd check on it the next day and there was no deer and open water!

Right here, down this lake about a mile and a half—it's called Seymours Point—in another week or ten days, half of that will be open, and there are places that are like that all over the place. And you remember them and you know where to stay away from.

Another thing I'd like to touch on. Before your time flying, we know, from the stories that Dick Folsom, Mac, and George told, about a lot of early airplane wrecks. And in many cases, the things learned from them became information for us so we wouldn't make the same mistakes. For instance, I'm thinking about an accident that occurred on Braley Brook. Do you know about that?

That happened the year I was born. That was a guy named Bill Szabo, who was a trapper. He lived around the Island Falls area and had a Cub. And he didn't show up for a few days, so they started searching. They found the airplane on Braley Brook. I've found the place where that happened. There's only one place where you could possibly land an airplane there, and even today you can tell where that was. Well, when they found the airplane, it was up on one wing a little bit. What happened, according to Mac—he told me the story quite a few different times—was that he landed, and as he landed it was pretty much glare ice. The wind started to drift him. It's pretty narrow and there's dry-ki there. He got out of the Cub when it was still moving and they used to kind of drag their feet along the ski and hang on to the wing strut. Well, as he was getting out, a ski hit a piece of dry-ki frozen in the ice, pitched him forward, and he went through the prop. The prop hit his head and of course killed him instantly, broke the prop, and there he lay. Well, they went a couple days and it snowed, eight or ten inches. And finally, somebody found the airplane. Mac and George went over in a Cruiser, and they couldn't land there—big difference between a J-3 and a PA-12. So they landed on Coffeelos and they snowshoed over, five miles. They got there, found the body, and they cut some of the trees out so they could go in with the Cruiser. And they went in and that's how they flew the body out. Couple days later, they did some repairs on the Cub, got a prop, and flew that out of there. Bill Szabo…

Listen, Elmer Wilson told me this story. A couple of years before that, he'd trap all over the north country. He built a camp on First Musquacook, and that camp was there until about fifteen years ago. Well, he was trapping and he landed on Chesuncook Pond. No, take that back. He landed on Little Churchill, which is up a little to the west of Churchill Dam. He damaged the airplane pretty bad on the landing. I think it was late '40s. And there was no flying that airplane out. He took the airplane, he strapped the wings to the fuselage, made some brackets. There was a lumber camp right on that little pond. That was before snowmobiles; there was nothing around. They had an old tote road [trail for teams of horses] that went down towards Churchill Dam—'course that's where the workers were coming from. And he taxied that airplane on skis from that pond down to Churchill Dam, down Churchill, down Big Eagle, across at Locke Dam, and down to Telos.

Incredible!

Unbelievable! He told me about it, and I've looked at that many times. That's a hell of a ride in a snowmobile. And that guy did [it] in an airplane. He had a couple of old mattresses that they tied to the fuselage and they tied the wings. He had to damage it some to get it out of there. But that was the only way to get it out of there.

I also was told by Dick Folsom that in that particular case, when they looked inside the airplane cabin, there was still one mag [engine ignition system] on.

What Mac told me was that the mag wasn't on. On a Cub, the mags are overhead on your left. He'd shut it off, but one mag was hot [not grounded]. So it never shut off. And that's what ended up getting him.

How about the Duck Cove crash?

All I've heard of that is what happened, and I've seen the pictures. That was pretty tragic. The young fella, you know, he did not have a lot of experience, and certainly a lot of the people flying into the back-country probably shouldn't have even been flying that day. That was really bad weather.

I remember Dick told me that the guy had actually come to him and wanted to be signed off on skis, since he was still a student. And Dick flew with him. He's a pretty honest guy, and he told the guy, "I can't sign you off on skis; you can't even fly an airplane." But at that time, there was almost no enforcement, so those guys were just going up and flying and nothing ever happened.

There were guys who were private pilots who ran flying services. And some other accidents here, a lot of the accidents, were before my time. But in Desolation Pond, there were two different accidents that happened there. One of them was a fatal, and parts of the wreckage are still there. And the second accident, I remember there was some fellas in a 185, the 260-horse version. Dana Toothaker, the warden pilot, was right on the lake when that happened. They tried twice to take off, like to the west-southwest. And the third time, they got off the water but ended up in a beaver flowage. And they could have done two trips and they wouldn't've had an issue. And that would not have happened. Over the years, different accidents, right there at Churchill Dam in the early days, there were two different airplanes. There's supposedly a wrecked Cub in the bottom of the lake there off the island on Churchill, somebody from Waterville. And then Johnson Pond—there were a couple of accidents there over the years. Some of the people going in there weren't flying-service people, not people who flew every day. It is not always the best place to go!

Do you know anything about a big search for the guys in the winter up toward Libby's and Munsungun?

That was before my time. I was a kid when that happened. I know quite a lot about it. My father sent two airplanes on that search. He remembers that at one time, right there where Libby's is, there were fourteen airplanes on the ice. It was two fellas from Fort Fairfield. They took off and were in a Luscombe. And they were headed to Millinocket Lake to stay with the dam caretaker. They buzzed by him that day. He saw them, but they kept right on going and he never heard the airplane again. Well, years later, I want to say fourteen or fifteen years, Sewall had a surveyor up there. The wind picked up and he couldn't use his canoe, so he parked it over by where Reed Brook comes in. And he started down along the shore going to the outlet of Munsungun, to where the little camp is down there, and he came across that wreckage. And he didn't know—he knew he found a wrecked airplane, but he didn't know what it was. Well, he got out, they had a Maine Forestry telephone line nearby, and they called out. Well, the next day Mac flew in there with a bunch of different people and there was those two fellas. I've seen the pictures of it and stuff like fifteen years later.

The other person that is sort of legendary in flying circles, just because he was so unorthodox, was Jasper Haynes. Did you interact with him at all?

My father did and Mac did. 'Course everyone has heard the Jasper stories. When you hang around Millinocket, you hear a lot of things. But I remember Mac telling me [that] in the early '50s he was over in that Daaquam, Saint Pamphile country, and back then you weren't going to see another airplane there because hardly anyone flew there. And Mac said he looked off towards Ross Lake and he saw an airplane that he thought was military. He couldn't tell what it was. When he got closer, he saw it was a Cub, and what happened was Jasper was headed out somewhere and had two bundles of rolled-up beaver pelts tied to the wing struts, like where we tie snowshoes. From a distance, that looked a little different! And of course, that's where his accident happened, right there on the river.

Dick told me that when they finally inspected that airplane, somehow quite a weight of traps had slid way back in the tail.

Yeah, probably out of balance. That's happened on some other accidents. People wouldn't pay much attention. You know, a Cub or a Beaver or any airplane will haul a hell of a lot more than they're supposed to, as long as you get it in the right place. But if it's towards the tail, that doesn't work very well. Jasper's airplane...my father remembers him at Red Pine one year, and he spent the night in a camp there. Don't know what Dad was doing there. He was getting ready to go in the morning and it was cold. He was heating his airplane...and he'd take birch bark, roll it up pretty tight, set it on fire, and he'd use the flame of the birch bark. And the soot would be all over the cowling. That's how he would heat the cylinders. 'Course the old Continentals, I don't care if it's a 65 or 90, those start in tremendously cold weather, way colder than what you should start them without preheat.

The only trouble we had when we were trying to start them without preheat was if it was really cold and the thing fired once or twice and quit. Condensation would freeze on the spark plugs and it wasn't going to start again. You had to take the spark plugs out and thaw them.

You know, that would happen on the 180s and 185s, too. You hoped it was going to keep going because if it quit you were done! You had to take the plugs out.

Well, that's an interesting start. Now what I really want to do is talk about what the job of a flying warden really means and your memories of the different phases of it. Probably a lot of the public just figures that a flying warden takes off each morning and circles around the sky looking for violations or something. But that's far from the truth. They're actually involved in everything from support of the biologists, to searches, to what they call BOLOs ["Be on the look-out"] for someone missing or a vehicle missing. I thought what we would do is begin with one topic, talk about it, and then move on to various other responsibilities. A search, for instance, can last several days and you may never find the person. Let's start with various kinds of searches. The typical types of searches would be lost hunters, drownings, escapees, criminals, smugglers, lost vehicles, and where we're close to Canada, a lot of this stuff can be really international. So what are your memories from some of the searches for lost people?

Searches, like you said, there are all kinds of different searches. Most of the time you're looking for someone that wants to be found, and occasionally, you're looking for someone who doesn't want to be found. You know—an escapee—hard enough to find someone who wants to be found. So our biggest thing in my day was the fall. I can remember some mornings having seven different lost hunters. And every warden wanted you first. They'd call you one, two, or three in the morning, "Can you come over here first thing in the morning?" Well, you can't be to everyone first. One day, Jack McPhee and I were the only two pilots working, and between him and I we had thirteen lost hunters. So you can't head to everybody first. So maybe a kid is involved or somebody with a medical issue, you might go there first. After that, I'd take them as they go. I'm not going to fly by two searches to get to the third one. I've found as many as four different ones in a day. But today it's a different story. They'll go a full hunting season and may only have one or two searches. Back in the '70s and '80s and before that, the other guys, they had lost people every day. There wasn't roads everywhere, nobody had GPS's. Now someone that's lost, they'll call you on a satellite radio, but he doesn't know where he's at.

When you found somebody, which is the hopeful outcome, were they pretty good about getting into some kind of an opening and being obvious?

Most of them were. You know, they used to have a sayin"Find a space, show your face." You try to get out somewhere, even if it's just a beaver flowage. Get out where we can see you. If you build a fire, we're going to get you right away. But very few people build a fire. It happened, but it didn't happen often. I mean, you can't imagine how far people would travel. You'd get on a search—we had, like, point last seen—and I can remember having a guy who got lost off the north end of Baxter Park, off the road. It was the Sunday after hunting season and I worked that search. I finally found his tracks. Had to be him; there was nobody else around. He walked across Frost Pond, Frost Pond being just west of Grand Lake Matagamon. And there was an old tote road there that went along Webster Stream. Anyway, late in the day, right before dark, I started seeing puffs of smoke way up towards Webster. I thought someone was starting a skidder [woods tractor] up there. That's what it looked like. But I got to thinking, something ain't right there! So I headed up there and as I got there, there was some big puffs. And there was a road that went parallel to the southeast shore of Webster. He was between that road and the lake, but he didn't know that. He had come, by then, miles and miles and miles. There he was standing, and he was torching birch trees. So I went by him, and he knew I saw him. I waved the wings and hollered out the 185, "Stay there!" He didn't move, and it took fifty minutes to get someone over there. One of the wardens that wasn't even involved in the search, we got him in there. And I stayed, and by the time he got there it was darker than hell! And you could still see the trees. With them lit up, you could see the red embers. And I steered a warden in, and I had him stop right there and lower his window and holler. He wasn't two hundred feet from the road, but the guy didn't know it. And we saw stuff like that all the time.

So that's the way you guys generally did it. You would steer the warden in. 'Course, he knows the road network but he doesn't know where the guy is; this was before GPS.

Listen, I've found over two hundred people. Some were drowned, some were deceased. Most were just lost and we found them. Quite a few of them I found, I was able to pick them up. They would be on the shore or something. Even in the '70s, it was quite a lot different than it was in the '90s. All of a sudden, there were a lot more roads. And there were a few people we never found. A guy in Baxter Park, we never ever found him. Also, a guy on the upper St. John—I think he was a scaler [measures timber] for IP

[International Paper]. We had a big search. We had the helicopter. The river was freezing; it was tough. And we never found that guy. We figured he might have gone through the ice. So we weren't successful all the time, but most of the time we were.

One thing people may not realize is that even though today where there is a road network in a lot of the forest, it's very difficult for someone driving the roads to know exactly what you are seeing from the air. I think back to small forest fires, and we would be water-bombing the fire, but we would also be guiding the forester into the fire. And he had big red numbers on the top of his cab—205, say—so I could call on the radio to 205 and know it was the guy underneath me. And I could literally give him lefts and rights and then tell him to go a thousand feet into the woods. If he hadn't had that information, they would be forever trying to find a fire. It would have finally blown up and gotten big and then they could find it, but that's not what you want!

You know, you talk about that… I can think of so many searches where we found somebody where they would not have been found without a plane. We had a guy who was hunting just north of Allagash Lake, between Allagash and Caucomgomoc. And they found his car and he was gone and he never came back to the car. So we went there and there was a big storm coming the next day. We had two planes up there. And I got there and Jim Welch was there from Southern Maine, and he was on wheels and the weather was coming in. I told him to head home and he did. And you start to expand your search. You have the guys on the ground and they have a pretty good idea and you start to expand.

If you can imagine, I'm up north and west of Allagash quite a ways, and I look out towards Baker Lake in case he's up there because he's been gone a day and a half. And I happened to look—late in the day and it was overcast—and I see this orange thing. And I look and know it is at least four miles away. And there was our guy, and he was on this stream that flows into Baker. I think it was Baker Stream or something. And he was in this meadow. I see where he had broke through the ice a couple of times, water up to his knees. So we knew he was wet. Finally, I got him to look up at me, and I got him to stop. And I was able to get one of the wardens within about a quarter of a mile. And I told them, "Listen, the weather is coming in. I'll make a couple of passes from your pickup to where he is. When I get there, I'll gun the engine a couple of times. He is not very far." So I did that a couple of times. He had a walkie-talkie. And then I got out of the area. It seemed like it took half an hour, but was probably eight or nine minutes, and he had voice contact with the hunter. That guy ended up losing a couple of toes. We may never have found him even in the fall of the year because he was a long ways away from where he was supposed to be. He would never have made it, no question! But, that's what fluorescent orange does for you. In the days before orange, we would never have seen him. Fluorescent orange—I was working when that came in. You can't imagine how much easier that made the warden pilot job!

I can reflect on that. Really, those fluorescent hats look like light bulbs in the woods. You can see them. The thing about looking down at trees is that normally everything looks the same. When you see something out of place like that, it stands out like a sore thumb. Have you ever ended up finding people who are so chilled, and basically in shock, that they have lost common sense?

Oh, yeah. Listen, I had a guy up in that Dole Brook country. Charlie Davis was the warden that called and said we've got a guy missing, Canadian hunter. And I had the Super Cub then. This was the last week of hunting season. It had snowed a lot, and there was like fifteen inches of snow on the ground. And I flew into Fort

Kent for gas, and I knew I could get back to Greenville. But you know, on a search, you bring as much gas as you can. I get up there and everything is froze—the little ponds, you probably could have landed on some of them. So I get up on this ridge where the wardens were and I'm looking, and the ice was building slowly on the plane struts and windshield. A Cub will haul quite a load of ice, but you have to have a place to go and I didn't. I come around the ridge and there was an old sugar camp [to make maple syrup] there. So I look down and there's a guy standing there, his gun on his arm, cradled. He's standing, but he don't even wave, so the first thing I think of is that he's a searcher. I came back around, and then I realized that it was our guy. No hunter orange. Been overnight in the woods and he didn't even wave! So I come back around and came pretty close—the top of the ridge was in the fog. You didn't have more than a couple of hundred feet to play with. You had ice on the windshield, so you could only see out of the sides, and I got Charlie to walk in to the guy. So I headed out and landed on Moosehead somewhere to throw in some gas. I didn't have enough to get to Greenville. But I had twenty-four gallons in cans. That guy wasn't responding. 'Course he didn't have fluorescent orange on. He was a guy from Quebec. I can't imagine that you wouldn't wave because you had an illegal hat. They claimed he wasn't all that cold. You got to remember it was around thirty-two degrees.

Hey listen, I had a guy showed up missing—this was after I moved to Eagle Lake. He was in that Bean Pond country; that's over west of Caucomgomoc fifteen or twenty miles. That's not where we were searching for him. That's where we ended up finding him about a month later—not a month, three weeks. Anyway, we searched for days and we couldn't find the guy's car. So for all we knew he was headed out of the country somewhere, an older fellow. And the people around knew him. So I was on another search about three weeks later and it was the last of November. And I'm going along by the Russell Ponds on a real sunny day, and I happened to look out and something caught my eye, like a reflection. I didn't think nothing and I kinda turned. It was only a couple of miles away, and I see that there had been an active woods operation. And I looked down and there was this car. And right away I recognized the car. It was the old fella's car. It was over by this Beanpot Pond and I thought, Oh geez! What happened? This was three weeks earlier when he went missing. There was an old, old road on the USGS maps, and he was trying to go on that; that road was back of Caucomgomoc. You couldn't have gone on that with an ATV, that's how old it was. But he got started on it because they had cut wood there. Well, he'd come around this hill and he'd got stuck. And you could see where he'd tried to build fires alongside and everything. I remember I got Dave Sewall and one of the other wardens to head in there. And the guys that had been cutting wood there before, if they had got up on the wood pile, they could have seen the vehicle down below. They would have found him in another week or so because they were headed that way. But two weeks before, they weren't even close to there. So the wardens went in and 'course he was dead, died right in the car. His dog was still alive! After three weeks, you can imagine what kind of a mess that was. They would have found him, like I say. So you're always looking.

I can remember when they told about Andy finding that PA-12 on the back side of Katahdin in Baxter Park. It was a gentleman from Millinocket. He'd been fishing up to Coffeelos Lake with another guy. They decided to come home and were headed by Doubletop Mountain, through that valley. The weather was bad…anyway, they never showed up. They searched and searched and found nothing for quite a few years. Andy was coming back from Shin Pond to Greenville after fish stocking. I think it was in the spring of the year. Anyway, Andy was up a ways headed for Greenville, and something like a flash came off the side of the mountain there. He went over, made a couple of circles, and there was what was left of the PA-12. It hadn't burned or anything. And right away he knew what it was. So they went in the next day, and it was that airplane that had been missing out of Millinocket for years. So you never know what you're going to see. Look at that airplane of Folsom's we never found.

One thing people who don't fly may not realize is that you can fly over the same place over and over again, and all of a sudden the sun is just right, your angle is just right, and something flashes. It looks just like a flashbulb, but it could be, say, the window of a plane, or a canoe. I found a canoe one time underneath a beaver house on shore, and all that was exposed of that canoe wasn't the size of a dinner plate. It flashed at me.

Now how about the folks that didn't make it, people who were frozen in the woods? A lot of people who hunt Northern Maine really aren't in shape to be going through the woods, and they end up having heart attacks and freeze. Were you on a lot of the searches that ended up with someone who was frozen?

Yeah, I found an airplane, right around 1980, a red T-Craft. The guys, two of them, were headed up to Big Eagle Lake ice fishing on a Sunday. We never heard they were missing until Wednesday. A close friend of mine called me and said, "Listen, keep this under your hat, So-and-so is missing, and the family hasn't called because he didn't have a pilot's license." So a big ice storm came in right about then, and then the next day the Civil Air Patrol called because they kinda take charge in airplane searches. But we knew our woods! So they called and wanted me to head up around Shin Pond. I said, "Listen, you can send some of your guys over there; I'm going to do our thing. I'm going to start where he left from, and I'm going to fly up generally where he would have gone." They were okay with that. We already checked with the Allagash Waterway folks, and that airplane had never got there. They had a ranger [who'd] been on Big Eagle and Churchill, and he had never seen that airplane. Okay, so we knew something was wrong. I really thought they were going to be on a pond somewhere, you know, broke down. So I leave early in the morning. I had Dave Priest junior, the local warden in Milford, a good friend of mine. He was in the backseat as an observer. Dave is one of those guys who always would click his watch. He'd say, "We've been seven minutes." So we take off in Old Town and we fly three minutes to the little airstrip where they left from. And I'm going to turn right there and head up-country. I make the turn and Dave says, "Hey, look!" He points down and there was the T-Craft off the end of the airstrip right nose down. Tail right straight up in the air. So we come around and land and taxi within two hundred feet of it. And we go there and, you know, when they're like that... I get there and the pilot, he was half out of the airplane. He was dead—just looking at him, you know, color... By then, this was a Thursday in February. Dave said, "Hey, this guy's alive!" By geez, the passenger was alive! He was sitting face down, and all he could move was his right hand. And he'd scratched so much on the ceiling that there was blood marks from his fingernails. He's sitting there that way for four days. He could hear the kids over at the little school across the road. The lady that lived right there, she didn't know there was an airplane missing. She knew a plane had left. When all the commotion started, when they came with the ambulance and Jaws of Life, she looked out the bathroom window and she could see the tail of the airplane. So what happened there, a lot of sad things. The family didn't call because of the circumstances. That Sunday that the airplane left, I was teaching Bill Varney to fly a Piper Saratoga.

We went right by there and I pointed down and said, "That's So-and-So's pickup there. His airplane's gone; he must be up-country." If I'd looked down at the other end of the airstrip, I would have seen the tail. No idea! But that guy—years went by and I never saw that guy. But one day I'm gassing up in Old Town and this guy pulls up and he doesn't get out. He said, "You don't know who I am." Right then I knew who he was. He had lost fingers and so forth...

The other thing that quite often happened during hunting season is that right at the end of the season, everybody was still on floats. And you were still dealing with ice problems, in terms of small ponds that were frozen, in terms of the plane water rudder being frozen down, and that kind of thing. But one of the things that happened from time to time is they would finally find a hunter who was deceased and who was frozen. And because of their shape, there was absolutely no way to get them into an airplane, so they were lashed on the outside. Were you ever involved in that?

I've done that. We had... That pond right behind Horseshoe that has a little camp on it, is it Mountain? Anyway, just two or three miles from Horseshoe there was a small pond and a guy died there. We went in to get the body and tied him to the float. It was the only thing to do. He wasn't froze but he was stiff. When I come to work, the fall I come to work, it was about mid-November, and George called me and said, "There's an issue down on Alligator Lake and Andy's off. Can you take care of it? There's a sporting camp on the island." So I went down. Well, the old fella had died. There was no roads in there; there still isn't. So we got permission. I knew him; he was a close friend of Andy's. I flew him to Old Town and they met us with the hearse. Six or seven years later, I got a call. They said they hadn't heard from the widow. They were old folks and she was still living there. Went in there and met a warden. He came in by sled and I flew in. She had died of a heart attack. She was on the floor, froze. So I flew her out.

They may be the only couple where both were flown out.

Oh, it's not really a search; something, you know, that happens. I know Jack has done it a few times. I've flown quite a lot of deceased people. And you don't just do it. You have to get permission from the medical examiner's office. It depends on what it is and the circumstances. Lot of times, just what we tell them on the radio would be enough. If there was any issue, then we have to go get the medical examiner.

Now, let's move on to drownings. There are certainly enough of them. People are out in canoes in the high winds and the early spring going; where it's cold, they don't last very long in the water. One thing I've always been curious about is, when you're called to a drowning, how often is it that you can actually look down through the water and see a victim?

A lot more than you would think. The glare is a big factor, and of course the quality of the water, the visibility. I've seen people in fifteen or eighteen feet of water.

A pond like Rainbow.

Oh, yeah! And then you might move over a little bit, and all of a sudden the angle is different and you don't see him. I can think of two or three dozen drowning victims that I've found. First thing I saw was the hands because of the white. Found a guy over on the river in Fort Kent, and I could see him only in a certain way. Then I'd come around and I wouldn't see him, then I'd catch him again. I steered a warden in. He jumped in the water up to his waist and never saw him. I walked him within five or six feet and said, "Hey, right in front of you." Then I'd lose sight of him. When you come around, you only see him for a couple of seconds. I had a guy in Van Buren I found; he'd jumped off the bridge and he floated. The reason he floated is he had nylon clothing on, some of it, and he had an air pocket. And when I saw his bald head, he was floating. Never had seen that before. When they drown, they go down, unless something...

I found a lot of people, and I mean, I've found them after they come up six or seven days later, depending on how warm the water was. The wardens who are on the boats, they don't share that view. They can't see them. I had a guy... We talked about the place in Abol where you have big water on the stream right above Abol Bridge. Right there in that long stretch, there was a car parked for two or three days. They wondered where the guy was, you know. Geez, they went down to the shore of the river and they found his clothes there. Well, what happened was this guy was growing marijuana on the other side of the river and he'd swim across. So they called a warden, figured maybe he was lost. And I came in—I had Leonard Ritchie, warden supervisor—with me, and we made about the third pass up the river and it was just right. Maybe there was five or six feet of clear water, and you could see him right there in the water. Come back around and you wouldn't see him. So there was nobody there yet with boats and stuff. When we landed and taxied, we knew we were right over him. Took back off, got up and went down to Abol, and we got one of the hoe rakes, like you dig potatoes. We taped a piece of two-by-two to it, we landed about where he was, let the airplane drift and never got nothing. We moved over, let the airplane drift and Leonard hooked him, grabbed him, pulled him, and we taxied down to shore. 'Course, technically, you're not supposed to, but you could put him up on the float, and we brought him down to the little beach right there. Yeah, I could see him real good. I've found a lot of drowned people like that. I mean, some you wouldn't see in four feet of water.

Then, as you've said before, sometimes you would have to suspend a search just because it had been so long. The water conditions and temperatures can be such that they surface much later on, right?

Oh, yeah, the temperature... some don't surface if they are in deep enough water. This past summer, we had a triple drowning here on Square Lake. And I got a call in June, I think. Got a call that we had a boat that was missing over there. Four guys in it, and they were going across Square Lake to fish the Eagle Lake thoroughfare. First thing in the morning, I left in the dark. It's not five minutes flying over there. I get over there and the sun was just breaking and the glare was terrible! But I wanted to find that boat. I figured they would be on shore somewhere, the motor had conked out. The wind was out of the north. I get way down... Square Lake is a rough lake. For a lake its size, it is unbelievable! And I get down there, and there's the boat, and I can see some guy standing there with a paddle. The boat didn't look right—well, it was full of water. So I landed and the waves were high. Good ol' PK [float manufacturer] floats! I drifted back to him, and I got out on the float and I yelled, "Hey, how you doing? Where are your buddies?" He said to me, "I'm the only one that made it." Oh my! So, I backed in and I got him in the airplane. He spent the whole night hanging on to the boat, from like five thirty. And when I got there, a little after five, he'd only been on shore half an hour. So he spent the whole night holding on to a boat that was swamped. I took him back to the other end of the lake where there is a road that comes in. The wardens were headed in there. And I pulled in out of the wind. And as I turn around and pull in to shore, a pickup drives in, two guys from Ellsworth. And I told them who I was and said I needed some help. The guy got out and we put him in the warm pickup.

Could he even move?

Yeah, he could walk, tough, tough, tough! Guy about...maybe he was thirty. Then he commenced to tell us the whole story. Well, that search went on for four or five days. Well, later that same day, I had a warden sergeant, Durwood Humphrey, a warden pilot with me, and we go down there. It's maybe two and a half or three hours after I found the survivor. And we were going along that shore—the wind was really howling. And we had a warden boat there looking along the shore because I had already found some debris from the boat, life jackets and a gas tank and stuff. Durwood looked out and there's a body on shore. So I said, "We'll call the guys in the boat and put them right on that." About then, I went to flip the switch to the transmitter, and the switch fell apart in my hand. Couldn't talk to nobody. We had to get that body. There was no way to go to shore with the airplane: all rocks, no beach. I said, "We're going to land." And we waved them over and the boat comes; she's under the wing and Woody jumped into the boat and they picked up the body. The only reason we can figure that that body was there and the other two weren't is that guy died of a heart attack and his lungs didn't fill with water.

So then they commenced the search, and the Maine Warden Service has a tremendous dive team, the best of equipment. We did that whole part of the lake. We took the survivor back out there later that afternoon. We figured about where they upset and which way the wind was. And they started with sonar looking. And about the third day, they located one body. You could see him right on the bottom, on the scan. And when they cover an area, after they were done with that area, say a day's work, they would be ninety-five percent sure he wasn't in there. They always go beyond the grid, turn and then come back. And they weren't seeing that far, but they were seeing pretty good. Then, about another day later, we found the other guy, maybe three-quarters of a mile further. 'Course, they all didn't let go at the same time. The first guy we found along the shore, he was the last one to let go, right before dark. The other two were right where they let go.

Well, anyway, a lot of them type searches. And I mean everywhere. You got to remember, in the '50s and '60s, they didn't have a dive team, okay? They'd use grapples. And even when I came to work in the '70s, a lot of places where sportsmen drowned there were no roads to get in there. So we flew the divers. And when you picked up two divers and their gear, all those weights and belts, you had a 185 load then. They'd all have four or five tanks. So if you had two divers, you had ten tanks. That's where the Beaver shined up. We had built racks—I mean, we could pick up five divers and all their gear in that old Beaver. That's when I enjoyed it, that and fish stocking.

So, the protocol is for the Beaver to stay with the divers?

Like this search last year, I always stayed there. You never know when something is going to happen. And it's good to do that for the family. You know: you've got an airplane there. And every morning when I got there, I circled in case somebody had come up or I'd find something else. I mean, because if you aren't there you aren't going to find it.

One other thing that came to mind—actually it's something you've already covered—but with all the need to communicate with someone on the ground, like a disoriented hunter, did you guys ever try megaphones?

Oh, I still got one of the original ones. It worked pretty good on the Cessna. You had to point it backwards. You couldn't have a conversation or hear him, but you could say, "Next right." I walked a guy three miles when I had the Cub. 'Course the Cub, you could have done it without the megaphone! I had a guy down near Grand Lake Stream, put him onto two different roads. I walked him right out to the warden and his pickup and I said, "Right there where the old road comes in, he'll be there soon."

But no one ever mounted a speaker on the belly of the plane?

No. Well, the last helicopters we had, the 47s with the turbocharge, we had a PA system. And the helicopters are great. Look, I'm a helicopter pilot. They were not always practical because you would be going so far away: you'd get there and not have much gas, slower than death. But you could see better. But I'd go to my dying bed thinking I could do almost as good with a Super Cub. And in a Super Cub, I can see out of both sides.

That makes a big difference.

A Super Cub, for most of the warden work, is great. Ideally, you could have five or six pilots in Super Cubs and that would have been great. But there's a lot of things we do, like fish stocking, that they wouldn't work. Some of the best years I had were with the Cub. Tremendous airplane!

I think exactly the opposite is true of dropping water on fires. I did that with the Beaver. Even when you really get the hang of it, you still miss sometimes. And I watched guys who were really good with the bucket under the Huey, and they'd stop and the bucket would swing, and they would drop the water right when it swung back. And that water would go right down the tree!

Got to remember, I was the first guy in the State of Maine to drop water with a bucket. We didn't have Hueys. We had a Bell 47 with a sixty-gallon bucket. But it was so effective, especially on brush fires Down East. And you could pick up water… I can remember some places doing six or seven loads to a Beaver load because they had to go to a lake big enough for a pickup. I could pick up if the water was two and a half feet deep. Some of the things I did back then, I don't know if I'd do it today. Places we'd pick up water, oh, oh, oh! But, it's a hell of a tool. The problem we had eventually was that we just couldn't afford it. You know, it was very expensive. We'd keep one up here for part of the year. Some things it was awesome with. But if you headed from here to Daaquam, you knew you had to have fuel at Clayton Lake. And when you headed from Clayton Lake to Daaquam, you knew you could only stay twenty minutes and have to come back.

Didn't you also start to have trouble getting rotor blades?

Yeah, you couldn't get them military surplus anymore. And they had to start buying it, and the stuff was expensive! Some things are timed out on a helicopter. When they get so much time, you don't rebuild it, you throw it away and get a new one, like the tail rotor. We had good years. I flew it quite a bit. Dana and I were helicopter pilots. But there a lot of things… Sometimes you can pick up a lost person if you can get in. I remember a Maine Forestry helicopter helping us on a search. They found the guy on a pond and they couldn't pick him up. I landed in the Cub and hauled the guy out. It balanced out. It's still a hell of a tool when you can afford it.

Let's talk about people that didn't want to be found. Escapees, stolen vehicles. Do you remember any incidents looking for escapees?

Yeah, quite a few different times. I can remember looking for a guy in Ashland. He had escaped from the county jail in Houlton. And

I don't know that he was a really bad guy, but he was in jail for a reason! And someone had seen him between Ashland and Portage. So I get over there, and I'm starting into this field where they'd seen him, and I'm looking. He's away from the road, and we think he's still there because by the time I get there it was two hours later. When I get beyond the field, I'm starting to turn back, I look back and I see him running across the corner of the field. I've had that happen a couple of different times. But, like you say, they're not waving! They know what you're there for. People would escape or leave a nursing home that had issues with Alzheimer's or whatever. I've found two or three of them. I remember a lady at Blue Hill and I knew some of her family. Just happened when I got there, and I found out who she was. So there were some blueberry barrens behind the place that she left from. And I'm looking and I turn around and look backwards, and this lady was hidden underneath a piece of old farm equipment. And she knew enough so the minute I was away from her, I see her get out and move to another piece of equipment. We did a lot of that. I found people who weren't helping you; it's just that they had medical issues. They didn't know enough to help you out.

I remember at one time, and I can't remember the details, but it was down along all the fingers of land that stick out on the coast, like Bailey Island. There was a lady that was missing from Central Maine and they knew what vehicle she was in. And they thought she would go to the sea and look out because she was having problems. And we flew along those fingers, but we never did find the vehicle. Someone else found it and there she was, looking out at the ocean.

Yeah, all kinds.

Now, how about on the Canadian border, people attempting to come across, evade the customs inspection. Did you get involved with that?

Yeah, we had issues. Most of the stuff that we worked on at the border was people coming in to Maine to hunt and fish illegally. There was a big thing going on in the '70s. I mean, guys were setting up camps on the Maine side. Because you got to remember that one side of the border is farm country and the other side is all woods. And they automatically figured there were more moose on the Maine side. The Canadian moose season—we didn't have a moose season—that was a big event! And we would find camps that were set up with propane and everything into Maine two or three miles, you know. And so we'd work that quite a bit. And we had a ground team; we had a regular operation going there! And we'd bring the helicopter and leave it there for a week. You had to be careful where you left it. You didn't want anybody tampering with it. The Canadian border, Quebec border, you can't imagine even today the amount of shacks on the border. Some of them have satellite dishes. They are right in the middle, they are right so their nose is almost on the American side. All that is perfectly legal!

Do you think the intent is to take a shot from there and then pull the animal back?

The aim was to draw them into the boundary. They would go and walk in on the American side. You know, we've caught them. We'd catch some over there. They just figured it was going to be better over there. Truth be known, it was probably better where they were. But that was a big thing. We worked that hard for quite a few years. And then fishing—probably there were more Canadians that crossed into Maine fishing than there was hunting. All these brooks, they'd go in to them in Maine because there was shade and better habitat for fish. It wasn't unusual to catch people three miles in. I can remember one of the biggest cases we ever got that happened in 1976. I was coming back from Saint Aurelie, and I was over in that Depot Lake country late in the day, maybe five o'clock. And I'm up a couple of thousand feet going along. And I happened to look down and there was a little bit of a dead-water in a stream about three miles from the border. And I look and I just saw something and it was a guy that was fly casting. And I saw the line. The sun hit it just right, I knew what I saw. I turned right away because if I flew over, that would have been a giveaway. I was maybe a mile away. I kept on going and called Ron Cyr and Jimmy Dumond, my cousin; they were about six or seven miles away. I told them there was a guy fishing three miles in. There was an old tote road that came in right where those guys had crossed the border and they went to that spot, then they walked in and met them. And they had one hundred and seventy-six trout, and half of them were under six inches. But there, you live and learn like everything else. If I'd made one circle over them, it would not have worked. You learn that stuff after a while.

In all the time you flew looking for violations like that, did anyone ever fire at the plane?

Not that I know of. I can tell you, when we had the first Bell helicopters, with the Franklin engines…they weren't even as good as a Super Cub. Charlie Davis and I were flying the boundary someplace between Point of Maine and Daaquam. That's where the border stops being a straight line and becomes the river. And we go along, and 'course there are these camps on the Canadian side, perfectly legal, they can have a gun. We were maybe two hundred feet high, at most, following the river. And we come along and look down, and there's a guy standing in front of a camp and he has a rifle with a scope that looked to be this big. And he had it on us! 'Course I know he's just looking through the scope. I'd been in Vietnam a few years before that where if people pointed guns it was the real thing!

All right, let's go on to a really valuable function of flying wardens, that is, support for the biologists. And we have deer, moose, bear, and fisheries. Let's start with fisheries. I know you guys did a lot of stocking.

Well, we did fish stocking, which is a big deal. Even today, it's still a big deal.

You might explain how you do it.

Well, funny how it started, with the Gullwing. They built a tank that went inside it, had like a piece of stovepipe come out the belly. And they had somebody in the back of it that dumped the fish. The fish were in big stainless steel milk containers, and they'd dump them into the pipe and they came out the bottom. Then they got to where they had a valve. They even had a tank that went in the back of the Cruiser. It wasn't big, but they could haul fish.

Did they feel they had a high mortality rate of the fish early on?

You know, they really never got into checking that. They knew you could do it because in places in Canada, they'd done it. And they found out you can't air-drop togue or salmon. But you think about it, you shouldn't have to air-drop togue or salmon because the waters you are going to put them in are all big enough to land in. They found out with brook trout that most of them were going to make it. They found out later that the success rate is ninety-nine percent. As long as you put them into the water, but if they have to walk the last quarter mile to the pond they are not going to do too well!

Was this before you oxygenated or after?

The oxygen came about, I want to say, in about '76 or seven. That was the biggest improvement that ever happened because you could haul way more fish and the fish were healthier. I mean, literally you could have picked up a load of fish here and took them to Fryeburg. And it didn't take a lot of oxygen. And then they started doing it on the trucks. They had tanks and pumped in oxygen. And I remember some years, when it got warm, it [oxygen] saved a load of fish. And you hauled a long ways. I remember taking togue from Greenville and going up to Musquacook. Now, with oxygen you could do that. We don't [anymore] because there're roads and you can pick them up closer. But in the old days, a lot of times we'd drop the fish and a lot of them didn't make it. When oxygen came about, that took care of all that. And we experimented before the oxygen: we'd put funnels on the front of the tank and that would send air in. It was a little better, but nothing like oxygen. The fish stocking really got good when they designed the tanks that mount on top of the floats. Then you could take two ponds on one trip, where before, with the tank inside the cabin, you had fish for just one pond. When you had two tanks, if you had a small pond, you could take one for one pond and one for the other. On the Beaver, we had four tanks. You just had to write and remember which pond was for which tank. I did most of that. I flew the Beaver; that wasn't even a load for that plane. You'd take forty gallons of water in each tank. So it was a hundred sixty gallons of water. Yes, it's a load, but in the Beaver, no big deal!

And what did you find worked the best for height above the water and airspeed?

The secret to dropping fish is not how high—yeah, you don't want to drop them from a thousand feet up. The secret is your forward speed. When the fish hit the water, they want to go straight down. So if you come over a pond, and you get down thirty feet off the water, and you're doing sixty, it's too fast and they're going to bounce and you will lose some of them. So you wanted to be slowed up, flaps on, and give them a chance. We have put divers and biologists right in the drop zone so they could check everything. And splake, if you have to, you can air-drop splake. In a few ponds, we've done that. Even with trout, if the pond is big enough, I land; why not give them the benefit of the doubt? You know, it costs a lot of money to raise fish. And those tanks that we are using even today, that's a Howard Lambertson–built product. He built the mounts, he built the solenoids that released it. Those were starter solenoids from vehicles. He built everything right from scratch. He built the molds that they fiberglassed the tanks in, put the two halves together.

Does that tank open like a clamshell, or does it rotate?

It rotates. The top is open and it's made so the air will burble in there. We've actually let a couple of states copy it. They'd come over here, and Gordon Pease, who does all of our work now, he built two different sets of tanks for other states. And they flew their airplanes over and had the tanks installed here. Something that we started in Maine. And we flew fish with the Super Cubs. Years ago, they always had a Cub as a spare airplane. Great airplane to haul fish with! You can't haul quite as much, but it's nice and slow. It was a manual release. And when we got the Cubs there, in the late '70s, we would make the tanks out of Seabee [flying boat] wing-tip floats. So we'd buy those, they'd cut them open, they'd make the mounts. And we still have three sets of those. Matter of fact, I'm going to try to get them to put a set over at the museum in Greenville, in the Maine Fish and Game corner.

Let's say you're going to stock Bill Morris Pond, and you get there and there's a guy out there in a canoe fly casting.

Obviously, you try not to hit him. 'Course you got to remember, the first years, up to the early '80s, they only stocked in October. So there was no fishing. Now, they fish in October. So a small pond, you got to pay attention. Some ponds I can think of—you know, some ponds are hard to hit! But we got to where [we] could do it good. We got pretty good at it.

Say you're going over a pond fifteen hundred feet long. Obviously, this mechanism takes a while to work. So when would you punch it off—as you're coming toward the pond?

Just before you crossed the pond because it takes that long, and then the fish have to fall. I don't think I ever, ever dropped them short, even on the smallest of ponds. But occasionally, a few might end up in the bushes at the far end.

That's good for the otters!

Oh, listen, when we used to do Tumbledown [Pond, on top of a mountain], that's a tough pond even when the wind isn't blowing. But it's almost impossible going across it to get them all in there. And they take that into consideration. They put some extra fish in. We got to the point when as you got there you'd turn and put them on the right side because it would be a small amount of fish. And you'd make a turn and go more along the length of it and then drop off the mountain. It drops a long ways there. Same with Speck, you know that's up pretty high. There's a lot of ponds that are smaller than that, but they're not on the side of a mountain.

Let's talk about moose. Once the moose season started, they were very concerned about herd size and so forth. They had set quotas, and quite a bit of flying was done for census work. Tell us a little about that.

When I come to work we were doing a fair amount of deer yard surveys, and then, along in the '70s, they started thinking maybe we can have a limited moose season. So we were doing a moose census. And what we'd do, we'd take the state and draw lines and you'd fly from one side to the other on the lines. We did that quite a few years.

Did you have a biologist with you?

Yeah, you had somebody with you. They were counting. In a Cessna, you look at your side and somebody else has to look at the other side.

What time of year?

We'd do it in the winter, and we wanted a fairly sunny day. We didn't want it overcast because you can't see the tracks well. That lasted quite a while.

I can say from experience that their black coat against the white snow stands right out. And funny, how many times they choose a knob [small round mountain] and there would be four or five standing on the top of a little knob.

Hey, listen, on Doubletop Mountain, over by Baxter Park, on the back side of that and down a little ways, there's a big flat area. And I've counted thirty-seven moose there.

Just for reference, for people who aren't from here, there can be three or four feet of snow where they [the moose] are living.

But you know, I saw some this winter, about a month ago snow-mobiling, and you know, there was a lot of snow, but they get around. You know, getting back to moose, especially when I moved back to Eagle Lake in '85, and I'd fly the waterway a lot. It was not unusual to see a hundred to a hundred fifty moose a day. I can remember flying around Churchill and Big Eagle and counting fifty-some moose in the evening. They're everywhere—everywhere! And even here, to go show somebody a moose is no big deal. I get a bigger kick out of seeing a bear. And we used to do the deer yard survey. They still do it. That—you go where you think there're deer yards. More so than with moose, you need sun. And we did a lot of that way back. They still do some. But the deer yard survey was a big thing.

I happened to work on that, as well. They liked my Cub. I can remember doing the Indian Pond yard and having Bill Noble with me. We'd fly up and down repeatedly. 'Course, he was going to go in there later and spot-check his work. Based on the tracks we could see and the animals we spotted, he would estimate the number of animals. That was a huge yard over behind Moose Mountain. They had a record of it going back years and years. The thing that was most shocking to me was when we did the Moose River yard, which was over near Spencer Lake. The highest historical counts there had been about a hundred and fifty animals. The year we did it, we spent a lot of time and he was satisfied there wasn't a deer in it.

The biggest deer yard in the state of Maine in the '60s was up on the Little Black River near Allagash, hundreds and hundreds of deer. Now, right in Allagash village, there're probably a hundred or hundred fifty deer, but that's all there are for miles and miles. You know, getting back to the surveys, a lot of what we did was out of Old Town over east, and a lot of it was up here. In the winter, we'd do surveys, we'd count the number of fishermen, then after a while they went to counting parties. For example, last year I would do the Allagash Waterway chain of lakes, and I would do the ones going out of here. So you'd do maybe ten or twelve waters—you do Millimagassett and stuff. It took an hour and a half, maybe.

You know how many people it would take to do that in a day by snowmobile?

Occasionally, you'd do creel census. You'd take a biologist and stop on certain lakes. That was a lot of good information. So what the biologists do now up here…like last winter they worked the waterway. They'd go to Big Eagle for two or three days. One of them would work on Big Eagle and one would go over and do Spider, Pleasant, and Clear Lake. That's an all-day deal! They taught the fishermen to take scale samples and stuff. So we've done that ever since day one, you know.

How about the radio telemetry of animals that have been collared?

Funny you asked that because right now, as we're talking here, there're fish swimming in this lake that they put a little transmitter on last fall. And I see the biologist over here three weeks ago with an antenna on the back of his sled, and they are tracking the fish in the winter.

How about bear dens?

I did bears forever. When they first got into the bear project, tagging them, that started over west of Ashland. And we'd track them during the summer and fall. Then we'd find out where they denned. And now with the GPS's and stuff, when you're tracking a bear, any animal, you can get him in an area the size of this lawn like nothing. I mean, quite often with bears, I'd see the bear. That's not what you're looking for, but you'd see him. When that started, I was in on the beginning of it. Jack and I both, we did a lot of tracking. That's where a Super Cub shines. I mean, we've done it in a Cessna—nothing like a Cub! And then we tracked—remember the second caribou batch they let go? Jack found one over around

Van Buren somewhere. The first tracking of anything I ever did was woodcock, down in Old Town. They put a little transmitter on them and we'd do woodcock. They've got a project going right now with spruce partridge. They've got a bunch and they track them. Very, very effective.

The fish is something they started a few years ago. When I worked for them three years ago, for four or five months, they had a bunch of salmon that they'd stocked below Fort Kent. I found one of them that went through the fishway between Fort Fairfield and Perth and was down the Saint John about eight miles into Canada. It's amazing! And some of them got caught. They found a brook trout last year late in the fall way down by the Square Lake thoroughfare. They kept getting a signal in this cove. The signal was so accurate they knew that it was near the shore, which didn't make sense. They found the carcass of the trout where a mink had got it, drug it up, and eaten half of it then covered it up with branches.

Amazing! Now some of those tracking devices—let's just say it's a moose. If it stops for a long period of time, it knows that the animal is dead, right?

Gone, or like bear when they den, it goes to a different beep. Each animal has its own frequency. Listen, bears, we had them over here west of Ashland and down to Stacyville. I knew those bears on a first-name basis, sixteen of them! Then they started having some over in Bradford, west of Old Town, and they had some out in Washington County. They got a lot of information out of that. They did lynx. They tracked lynx and they had radio collars on them. There's not much they haven't put a radio collar on.

Just a little background on the lynx. It was debated in Maine for a long time whether we really had lynx. Then there were confirmed sightings, and this has to do with the Endangered Species Act. Eventually, there was no question that we had lynx. So then it became an issue of how many. And the same information pertains to mountain lions, which many people feel we have and many people feel we don't. Do you know people who are knowledgeable that really feel we do?

I don't know anybody that I consider knowledgeable that thinks so. I've had people state to me that there's no question what they saw. And hey, anything's possible! You know, when I came to work in '72 for the Maine Warden Service, we were just starting to hear about coyotes. Many people didn't believe that! The first coyote I saw, I had Leonard Ritchie with me, and it was on Mud Pond, just west of Chamberlain. And I see this thing and I mean, it was going. It must have been running thirty miles an hour. Couple days later, we're up to Chandler, where Libbys have a camp, and it was midwinter. Leonard said, "Hey, look at that!" So we come around and landed. There was a coyote deer kill, and that's the first one I'd ever seen. And it looked like they'd massacred something there, and they had. There was blood, and all that was left was a few little bones and a little bit of hide. Then coyotes changed. I flew all last winter, and on a good flying day, yeah, we'd see coyotes. I can remember in the late '70s and '80s, you'd go out and you'd see coyotes everywhere, deer kills everywhere.

I was living in the woods in the '70s, and I can tell you that the coyotes just exploded. We had ice fishermen on the lake, and Spencer had yellow perch, which everyone wanted to eliminate. So they would discard them on the ice. We'd be in the camps looking out, and just about the time it was getting dark, coyotes were coming out everywhere to feed on those. The other thing is, there were cliffs on both sides of the lake. So you'd be there at night and it sounded just like you were in Alaska with wolves. They'd call back and forth, you know. And another anecdotal thing which is interesting. There was a hermit's camp near our lodge, and the man kept pencil notes on the inside of the logs, interesting things he saw. One of them, dated 1946, said, "Wild dogs running on lake." That's seventy years ago!

You know, you talk about the perch that they leave. When I came to work, ice fishing was February and March. There was no January. And there was no limit on whitefish. And I can remember flying around Big Eagle, Churchill, and picking up whitefish. People would leave them, wouldn't eat them. I mean, I remember piles of whitefish—there'd be twenty of them. And then they went to an eight-fish limit, then they went to a three-fish limit and then you couldn't keep them. Now, you can keep one if it's a certain size, so that's how things change.

Now another thing that we get into are people who have no intention of following the law. And probably the top of the list would be poachers. And these are organized deals where you have a bunch of people killing as many deer as they can kill, many times trying to sell the deer. What do you remember about various sting operations?

You know, there are some people that just have to go do it. Some of them are probably worth a lot of money—they don't need it, they just have to do it. It's a disease. And I remember we had an operation for quite a few years. They still do a little of it. But for quite a few years, we did a lot of night flying. Somewhere there's always somebody, especially in the good deer country, that's Central Maine. But it's hard to be where they are at the right time. So we started to use the airplane. We thought we were going to set the world on fire! Night after night for four or five hours, but you found out there was no way! Once and awhile [sic] you would see somebody lighting a field. We did have a couple of good cases. We were flying that with two pilots, which I didn't agree with—needed another observer, but not two pilots. And then you'd tie up two pilots, and one was off the next day when you had a few lost hunters. We'd end up going. You'd fly until three in the morning, and then at six, you'd be back out. Towards the end of it, we were flying single pilot. If fact, the last few years I did it, I did it on floats.

I don't have any problem flying at night, like out of here. 'Course, back then we did not have GPS, you just knew where you were. You had to watch that Central Maine country because…I remember one time flying out of Old Town on wheels, and I'm over in that Waldo County country—Belfast, west of Belfast. And all of a sudden, I don't see the lights of Belfast anymore—what the hell! The fog was rolling in. You don't see it, and that's how you notice it. I called Orono and they called the Orono police department and they went to the airport and reported that it was fogged right in here. I saw the lights of Pittsfield and we went along over to Pittsfield. So that would happen; that's happened a few times. We did quite a bit of that night flying. I think what it did a lot is to let them know we were out there. 'Course, you'd fly pretty high.

I'll tell you an interesting sting operation that we did when they were bringing illegal bait fish into Maine. They were buying those in Massachusetts and Connecticut and we had undercover guys working that would follow the vehicle. Well at night, every vehicle looks the same. So they had a magnetic beanbag with a certain color light, and they'd stop the vehicle somewhere and throw that on the roof. Well, it's easy to follow a pink light! I followed a vehicle from Lawrence, Massachusetts, to Pittsfield in the Cub. It's hard to do that in a Cessna because you worked so hard to slow it up, coming around. In the Cub you could just set it at fifty or fifty-five and go. And we did a lot of that. I mean, that was amazing!

A lot of work was helping the State Police on drugs. [People] were bringing stuff in off boats. The big issue there was that they had a couple of pilots, but those guys had never done any of that, you know? It was pretty easy to pick up a vehicle in Camden and follow it into Bangor. When you looked at stuff all your life on the ground, a vehicle wasn't hard, but those guys had a hard time with that. We'd end up working for them and we'd fly their airplane. And I remember one time, they had information there was a boat coming in. That's when Loran was getting pretty popular, and they'd meet a bigger boat out there and pick up. Well, they had information and we'd pick up this boat at a certain port with the airplane. We'd be up at five thousand feet—they had no idea! And we'd follow it, and it would go quite a ways out to sea and stop. Well, we'd see what they were doing. They were making pickup points on their Loran. And I remember one time we were out there real high. I had a State Police detective with me. Everything was civilian clothes because you're trying to keep a low profile. And geez, here comes an amphib [type of airplane], and it went over a few times, wiggled the wings, and I followed it, but I couldn't get the N number 'cause that's when they'd gone to the three-inch registration numbers. But we did a lot of that.

One time, I'm out there in their [State Police] Skyhawk, had a guy with me, and we were out there a long ways. It was hazy, in the summer, and I could just barely see Vinalhaven. I said, "This is a long ways to be out here on wheels." He said, "You can swim, can't you?" I said, "I can't swim thirteen miles!" So the next time we go, he shows up with this rubber raft—you pull the cord and it inflates. "Don't put that in the back of the airplane!" If that let go in the airplane, we'd both be dead. It would crush us! We did a lot of work for them, and even now they do some. They got their own airplane, but theirs is mostly for traffic enforcement.

Now, in terms of what we call fish hogs—these would be people who are going to catch just as many fish as they can cram into a cooler.

Quite often, they exploit a pond that has a delicate fishery—in other words, there are not that many fish in it. So how do you guys, once the rumor mill starts…how do you set up to get them?

Well, there's a lot involved there. First you got to know when they are going to go, or when they tend to go. And you've got to have a little help, somebody has to talk, you know. I can remember different ponds and getting guys coming out. Cedar Lake up back of Millinocket, and it wasn't even open because it opened in late May. The guys come out of there and we got them. They had well over a hundred fish. We got kinda burned on that one. We got them for fishing a closed pond, but couldn't get them for over limit because there's no limit on a closed pond. So we thought we had them, and we had them, but…

So when someone commits a flagrant foul like that, does it ever jump into a class—as far as the law is concerned—where it becomes a felony?

I really don't know that. I don't think so for fishing. Maybe things have changed now, but I don't think so. I remember when you'd catch somebody over the limit, it was so much a fish and there were some pretty sizable fines! But you know, for it to work, you almost have to have someone talk. But most of them that do that, they've got to talk! That's why when they let the caribou go and they disappeared, people claimed that someone shot one over to Patten, got one in Stacyville. We didn't believe that because someone would have to talk about it.

I remember a fish hog incident in Little Enchanted Pond; that's a small pond. The biologists did a fish census on it, and they thought there were about six hundred trout in the whole pond. Then the wardens caught a guy with over a hundred trout. So that just shows you the damage that can be done to a tiny pond.

We had a pond not far from here just six or seven years ago that had big trout in it. You could get two-pound trout in it! And a couple of guys went in there two times in a weekend and they cleaned the pond!

And the last category of offenses is Sunday hunting. You can't hunt on Sunday in Maine, but a lot of people do because they like the place to themselves. So I know you guys fly a lot of Sunday patrols.

Sunday hunting was…my feeling was the main reason for no Sunday hunting was the safety aspect. If you allow it, there would be a lot of people out there on a Sunday. I can think of a couple of real good Sunday hunting cases. It happened; that's the way it was. I remember Leonard Ritchie and I flying over Grand Lake Seboeis, and there's a camp way up in the northeast corner. You couldn't even drive there. You went by boat or you went by airplane. We knew they were in there and flew in. We landed about two miles up the lake, kinda quiet, and we walked down the shore. Seemed like it took forever! We got up there and the camp was warm, nobody there. And we sat on the porch, and we sat for like three hours. First thing we know, here come two guys. Their guns were loaded, and we were sitting on the porch right in our red jackets. 'Course, they had agreed to be back for lunch at a certain time. "Sit down, guys, and don't say anything." Geez, [then] the other two guys come, you know. And I can remember flying over Harrow Lake high on a Sunday. Harrow is over just east of Churchill there; only one camp on the lake. And it's hard to get to; you can't drive to the camp. And I'm up high and I see the back end of a square-stern Grumman canoe pulled into the bushes. I could see a little bit of the motor across the lake. There was no question what was going on! You don't turn there, you don't even dip the wings, you just keep on going. And I call Jimmy Dumond, who was the warden there at the time, and he come over. When they got to the canoe, he was sitting there waiting. And they're not necessarily bad, bad guys; they just want to get all they can.

There's sort of a complicating factor in Central Maine. So many people have built little comfortable houses to hunt out of. When you look down at the fields, you'll see one little green building—looks like an outhouse—right after another. So from the air you can't tell if anyone is in it.

That's happening up here now. We've started to see people building little buildings.

On another subject—we already talked about the weather turning sour, and believe me, it can happen fast. I know in the early days it was normal to be stranded out for the night. What do you remember from those early guys about realizing that they were going to spend the night?

Even just speaking for myself, the first ten years, maybe. Quite a few times a year, you wouldn't get home. You might get stuck in Millinocket. I can remember getting stuck at the Umsaskis camp one night. Hey, it's nice to have a camp to go to. Now that you have a radio, you can tell them to call and tell your wife. In the early days, they didn't have that. They could have been missing two or three days. Mac Maheu was coming back from Waterville—that's when they were doing the maintenance in Waterville. That was in the late '40s, and he was flying a Cruiser. And he was coming right back here to the Fort Kent airport because Eagle wasn't froze; it was the first week in December. And Mac got just east of First Musquacook, and he lost oil pressure. He turned back, and he knew about that camp they had. They had just took over that camp—that was Bill Szabo's camp. So he had to put it down and the lake wasn't even froze. But along shore was froze, and he had to go somewhere with no oil pressure. He set it down along the shore pretty close, and just as it stopped, it went through and went over on its back. He got out—never got wet sitting on the tail. But he had to go to shore. He couldn't call anyone and let them know—no radios. He was a hundred yards from the camp. Finally, he got out and he was in the water up to his armpits. With his elbows, he broke that shell ice and went to shore. Got into the camp, there was everything he needed except kindling. There was cedar, but it wasn't split and there was no axe. He looked everything over and he was cold. Goes back, went in the water, reaches in the airplane, which is upside down. And he goes under and he can't find his axe. Then, first thing you know, he found it under the instrument panel. It went forward when the plane flipped. He went back, split kindling, got dry, spent the night. The next day, he went down to Squirrel Pond, halfway to Ashland, like, and spent the night. The next day he hit a road. And they knew he was missing because he hadn't showed up here. And Sleepy Atkins picked him up walking. That was the year they called Mac "the walking pilot." In three months, he went down three times. Actually, he went down four times, but one of them he didn't do much damage. He went down back of the Clear Lake camp, lost the engine, and never got a scratch on him. And then with the airplane that went through the ice, he went down back of Portage and put it into the trees. The thing never hit the ground completely! There was an old tote road, and they went in with one of the little bulldozers, made a sled, and they hauled the airplane out. That's when he came out with the saying that he had: "Did you ever climb down out of a tree that you never climbed up?" So there was three times in a short time.

Mac was a tremendous pilot; things just happen. That fall and winter, he was picking up some beaver traps. They caught a trapper in Pete's Pond, that's over beyond Allagash Falls, east of the Allagash four or five miles. There is an old camp there. So Mac went in there and he hit something and he busted a landing gear. Didn't hurt a thing, but you aren't going anywhere. The trapper they caught in there a couple of weeks earlier, 'course he caught a lot of beaver. There were carcasses everywhere. Well, the weather got bad and [Mac] was stuck in that camp. He spent two or three nights there and ate beaver meat. You know I've had it, pretty good. If they hadn't told me, I'd have never known. I had it in a stew at regional headquarters that Roger Auclair's wife made. So Mac is stuck in there. They had a pretty good idea because he told somebody he was going to pick up the traps. Well, the weather's real bad; George and them can't get out of Greenville. My father, he had two airplanes, had a guy working for him. They were going to try and get in there, but they can't leave the Allagash, that's how bad the weather was! The next morning at daylight, the third day, they went in and got Mac. He said, "What took you guys so long?" So you know, lucky there was camps everywhere. You know, when I came to work, I can name a whole bunch of camps. They're gone, they just burned them.

'Course the big difference today is that there's almost complete radio coverage everywhere.

There is. In the airplanes now, they have got that GPS that allows you can text on satellite. And they have the regular ones and they have the GPS that tracks the plane. You can send a text to Ashland headquarters and tell them you are stuck on Umsaskis. They'll text you back. Satellites!

And I'll tell you why that's important. I was in a 185 one time, and the passenger got violently sick. So I set down, and we'd been having trouble with that plane in terms of starting. Sure enough, when I went to start it, it wouldn't go. There wasn't much battery. So I was getting ready to spend the night. And the important thing is, I'm miss-

ing for hours and you can imagine how your wife feels! So to make a long story short, George Later found me and I got out. But with the radios today, it certainly comforts the family because they can know right away where you are and that everything's good. It's just a matter of when you can get back.

I've got a little radio now that DeLorme sells. It's called Spot. It's a satellite transmitter, and I can put down ten e-mails and ten telephone numbers, and if I press "everything's okay," that message will go to them. Or I can push another message that is "help." You press that button and it will go to the same people. You can also put the message the way you want it. It will show right where you are, Spider Lake or wherever. Then there's the SOS version. For that, you lift a little flap, so you can't accidentally press that. You want to make sure you need that because things happen.

About five years ago, maybe six, one of those Cirrus airplanes crashed over back of Saint Pamphile on the American side. They were flying in bad weather, ice. They iced up and they went down. The pilot got killed. The passenger didn't. He had the Spot and he knew how to use it. The pilot had showed him. He tripped that [SOS] and that saved his life. The wardens headed up in that country, and they were within maybe two hundred yards of it on snowshoes. There were no roads, and it was in a big bog area. About then the sky lit up! There was a C-130 [large military plane] circling, dropped a big flare, and in come one of those twin-rotor Canadian rescue helicopters. And in a manner [sic] of seven or eight minutes, they'd picked up that guy. That's how accurate that is. So they knew the guy was alive because he reset it and sent it again about twenty minutes later because he didn't know if it went out the first time. I got one, just one hundred and thirty bucks! I use it snowmobiling, but we didn't have that stuff back then.

One thing I want to talk about are the airplane incidents over time involving the Maine Warden Service. What's really amazing when we start to talk about this is that, I don't know—you might be able to hazard a guess—it wouldn't surprise me, since 1950, that the Warden Service has flown a hundred thousand hours. What do you think?

Yeah, since day one, yeah.

Counting all of the pilots together, in a normal year, how many hours do you fly?

Well now it is considerably less. It's a different ballgame, okay? Now, a good year would be three hundred, three hundred fifty hours each.

How many pilots?

Three.

So you're up to about a thousand total.

I can remember Dana and I, couple different months, flying one hundred and thirty hours. And I can remember Jack and I, when we had the Cubs, flying nine hundred. I can remember on average, when there was four of us, that in one year we flew a little over twenty-eight hundred hours.

So you can see, it's just an incredible number of hours. And over that time, there have really been only a handful of very serious airplane accidents. Which ones can you remember?

I can remember them all. I wasn't around for the first one, when George Townsend drowned in Maranacook Lake. George was one of the first five pilots. It was a brand-new Cessna 180 he had just gotten, and he was taking off. He had a biologist with him, and just as he pulled off the water and started upwards, near as they can tell, the passenger seat let go. They didn't have the safety stop then. They figured the guy grabbed the wheel. The plane just reared up.

And when that happens, the seats in a Cessna can go back a long way. And if the passenger slides back, number one, you get an abrupt change in center of gravity, and number two, no one can resist reaching for the wheel. And when they do, they can overpower the pilot. We actually had one in Greenville where it was the pilot seat that went back. And he went in.

Yeah, that was unfortunate, a brand-new airplane and they both drowned right there. Then the second accident was Dick Varney. When we lost Dick, I was working for the Maine Forestry, and I'd known him since I was a little kid. And he was right at Maranacook, about a hundred yards from that first accident, and he was flying the Bell helicopter, and that had the Franklin engine, the early ones. They iced up [carburetor clogging with ice] pretty easy. He was coming in, and just when he started to add power, that thing backfired, iced up, and it settled in the water. People watched it. Next, he kinda rolled, the rotor blades hit the water, broke the blades 'cause those were wooden blades. And they saw him get out and thrash around a little, and then he was gone. The tail of the helicopter was out of the water. He could have stood on the framework. Dick was scared of water! No one really knew that. I took his place—I come to work about a month later. And that winter, game wardens learned to swim. Learned to swim and at least know what to do if you get in that situation. It's sad, but that's what happened.

I had a good friend who fished for lobsters all his life. And he told me one time when he was in his nineties, "Jake, if I go over in my boat,

I can't swim but land is never far away. It's usually about five hundred feet straight down." So that's two of them.

The last pilot we lost...this week will make seven years—Daryl Gordon. He was over on the [Allagash] Waterway, and he helped a young warden with his sled, and they got it out. 'Course, February and March in this country, it snows almost every day. You know, snow squalls. You go four or five miles and the sun's out. Then, if you wait till five or six o'clock, the last hour of daylight, it clears right off north of the park. Well Daryl got headed home, and he never went very far. He got into some pretty heavy squalls. It looked to us, when they examined the GPS down at the FAA, that he decided to head into Clear Lake. And it was snowing pretty heavy, and we know so because twenty minutes before the accident, two people we know were fishing Clear Lake and they left by sled. They said you couldn't see any of the shore, it was snowing so hard. We had another retired warden who was over on Harrow Lake, which is four miles away, and the sun was out. He actually heard the engine rev. What happened is that [Daryl] got into snow and all of a sudden lost it. And in a matter of no time, like half a minute, you come over and there's a bunch of fir and spruce under you, and then all of a sudden you're over white. And he made a turn, and in just seconds he hit the ice hard. I went in to help Charlie and them get the wreckage out. That's what happened there. That's the only three fatalities. That's three too many, but it's the name of the game.

We've had a lot of minor accidents in the Warden Service, damaged stuff, you know. This airplane that went through the ice here this December, that wasn't the first time that a Warden Service airplane went through the ice. I think that was the fifth time since the '40s. Those things happen. And we damaged other airplanes. I remember Dick Varney landing on Frost Pond near Rip Dam and hitting a ledge. He flew the airplane back to Greenville and it was totaled. The fuselage was no good. I would say, if you never want to damage anything, just stay home and watch TV. Things are going to happen!

And especially on skis. People who haven't flown skis have no idea. It's not uncommon, for instance, to get a beautiful fluffy snow with no wind. And underneath that layer, say eight inches, are all these obstructions. Could be a frozen drift, could be a piece of dry-ki. So when you're coming in to land, you have to use common sense. By the same token, you're not going to see that frozen drift. I have hit those before and bitten my lip, the jolt is so hard. So you can see how it wouldn't be much of a stretch to pull a landing gear right off.

Ski flying can be the nicest under the right conditions, but it is the hardest on an airplane, way harder then floats. I've seen guys on Big Eagle get stuck in a Cub—two of them grab the tail and just start yarding around. They broke the landing gear. Got to remember that axle, when it has a tire, there's only a little of the tire touching the hot top. With a six-foot ski, the airplane turns, the ski doesn't!

We were talking last night about the same thing in the powerful airplanes, say a 185. If your skis are frozen to the ground, which is real common, it's awful tempting to apply a lot of power from that three-hundred-horse engine and work the rudders so you can snap the skis free. But what can happen is that you break the axle or spring landing gear. So you just learn, over a long period of time, to always think and take it easy, to do things the safe way.

Now, the other thing that I think is real important is this: we've talked about only a few serious malfunctions, and there's a reason for that—great mechanics. So let's talk about some of the mechanics that have been around and kept us safe.

Well, I'm going to tell you, Howard Lambertson—you knew Howard—he kept me alive! At one time, he was one of a handful of licensed FAA mechanics in the state of Maine. He used to work on my father's airplanes in the late '40s in Waterville when he worked for Marden Airways. Then Howard came to work for us. Our first base was not where it is now; it was where the Maine Forestry is now. That was the first Maine Fish and Game hangar. Then they built the new facility where it is now. And Howard was the face of operations; the guy was amazing. Just the nicest guy, and boy, he knew his stuff! I mean, he was around other talented guys. Him and Humpy, over to Folsom's, they were pretty tight. One of them would get a problem and he'd go see the other one: "What do you think of this?" I'm just so blessed that I lived during that time. Before Howard, they had someone doing a lot of their work, and I'll tell you who it was: Bud Aubin. And Bud would help them out. He had the little hangar up to the airport. And he did some of their work for a while. When Jack and I got the new Super Cubs in '78, about the first of July, Howard worked four or five days, got them on floats, got the radios installed, then Howard retired. Then we had a couple of other mechanics for a couple of years and they were really good. Then the Warden Service started contracting the work. We had some good people. The last twenty years, Gordon Pease has done all their work, [and] the guy is good. We were spoiled because we had the best equipment and the best people to take care of it. You got to remember, we all helped. What I learned there, working around people like Andy Stinson! Andy Stinson was and still is a tremendous aircraft mechanic. Good engine man, but a fabulous fabric and sheet-metal guy. There's nothing Andy couldn't do, but the same with George and Mac. Mac was a perfectionist! He built a house over to Wilson Pond—I mean, he just didn't drive nails, everything had a reason. My father thought he was the end of the world!

One thing that was different in the earlier days, I know because I spent so much time talking to Dick Folsom. They flew so many hours a day because the planes were small and slow and they couldn't take big loads. So it wasn't uncommon to turn around and work all night on mechanical things, get a little bit of sleep, and go out the next day.

Andy told me, when he worked for Dick—you got to remember Andy worked for Dick Folsom for, like, six years—he told me they'd bring in a Champ and do a top [cylinders] overhaul late in the afternoon and at dark, they'd test fly it. When I went to work for the Warden Service, we worked on the airplanes. I got to where I could spray floats, and I enjoyed doing that, but that's what we did. You didn't just sit there if you couldn't fly. It was different.

The other thing that's important is the airplanes can't always be perfect. And people with the experience of Howard, Dick, or Humpy could evaluate a cylinder and say, "It won't let you down, but next time we're slack, we'll change it." That's really important because you can't just tie these airplanes up all the time trying to make them factory perfect.

I can remember having an engine in the 185 the first year with the two-bladed prop and they vibrated. And you would never get a thousand hours out of the engine without topping it. There were different issues. They're better now. They have better engine mounts and smoother props. I remember having an engine that was burning almost a quart of oil an hour. The compression was good, and Howard said, "We're going to get another couple of hundred hours out of it, just keep putting oil in it." And it was fine, it just burned oil. And then you get another airplane like the one I flew the last year and a half, I'd add a quart about every twenty hours. We used to do our engine overhauls ourself, right in Greenville. We didn't get factory engines. Howard would rebuild them and we would help. We always had a spare engine.

Dick, of course, did the same thing. They did all kind of things that they don't do today. One example is the "rope trick," where you can remove a valve from the guide so you can ream the valve guide without removing the cylinder. You put rope inside the cylinder so you can use a valve compressor. I can remember Dick doing that to my plane because a valve was starting to stick. We simply reamed that valve guide, put the valve back in, and away I went. The modern system of fixing that is replacing the cylinder.

Another thing that is really important as far as the Warden Service is concerned, and I know this personally because I used to operate a sporting camp. The flying wardens make a habit of checking on people for good-will visits. And one of the reasons is that, if they have a smooth relationship with the sporting camp people, the camp owners are going to help them enforce the law. So what do you remember about that?

Yeah, enforce the law, and I can tell you different camps or flying services call and tell you, "Hey, we got people we're picking up; they'll be out Thursday at noon, you want to be here." The camps don't need people taking too much of anything, either. You know, generally the sporting camps—that was one of the great parts of my job. I just loved going to all the camps. Like Nugent's, Libby's… Libby's, I've been going in there since '72. The owner there now, Matt, he was a baby then. But I enjoyed that and I still enjoy it. And you know, sometimes you get a sandwich!

There's another part of that that people might not think of. We were twenty miles from a road, and every once in a while, we would have someone who was undesirable. You don't want to confront someone when you're all by yourself twenty miles from a road. That could go sour real quick! We had a radio, and more than once we called out on the radio to Connie and told her we had a problem. And guess what? A flying warden could divert.

You know, I remember a bad incident in Nugent's. Guys are in there ice fishing and 'course they're drinking and there was an assault. Somebody got hit pretty good and got hurt pretty bad. We had to fly guys in early in the morning, and they arrested people and it was quite a mess. But say they didn't have a radio to call somebody. They must have called Connie because the only radio they had was Folsom's. That was back in the '80s.

When Folsom's radio came in in the '70s, and they put a repeater on top of a mountain, all of a sudden all of the sporting camps could talk. And it was great for reservations, number one, and it was great when you needed the plane to bring in groceries. But for that emergency where someone got cut or if someone was unwelcome, it was a lifesaver. For an intruder, the other alternative is some kind of armed standoff, which is the last thing anyone wants.

We had issues, a couple I can remember. I remember Connie calling that someone had got hurt at Chesuncook Village. Burt had called out to let her know. Well, Folsom's didn't have a pilot free, and my airplane was in the hangar with a cylinder off it. Connie says, "We've got a couple of airplanes here." So I went over and got a 180 and headed to Chesuncook, to Burt and Maggie's. They had a guy broke his ankle. His ankle was more than ninety degrees to the right. And we got him into the airplane and brought him back to Greenville. If they hadn't had a radio, that would have been a major deal. And that would happen; there's always someone getting hurt. Or somebody hasn't heard from someone else and they are worried. A couple times over the years, we checked on overdue airplanes. I remember going to Ross Lake to check on one, and the plane was stuck in the slush. Another time, guys from Pittsfield—I get there, and there sat the plane with the landing gear twisted on it. There are enough camps that they would break into someone's camp, but those things happened.

Now, I've thought of quite a few topics; maybe you have some, Gary?

We have talked about winter flying on skis but not winter float flying in late November. You know, I've seen Folsom's with a lot of late hunters, and they would come in and they'd have four deer tied on the Beaver, two on each side. Ice—the deer were just a glob of ice from takeoff and landing. And I can remember flying four or five days in a row, and the ice on the floats didn't melt. It can build up, and that's a lot of extra weight. And we used to put on winter float cables on the 185s. When I came to work, I'd never heard of it. Another Howard Lambertson design. He'd take off all the pulleys on the deck and run the cable direct from the post on the water rudder direct up to the fuselage and back. It didn't steer near as good, but it steered a lot better than something that was all froze. And we'd put those on first week in November and fly all of hunting season with the winter cables, and we'd leave the water rudders down. And I remember the years of flying the Cub, you're closer to the water. The ice was so built up on the floats that you had to hang on getting in and out.

Tell them what it's like to try and get a passenger into the plane that has ice-covered floats.

Well, when their foot leaves the dock, which may be covered with ice, they've got to be hanging on. You got all you can do to hold on yourself, and you're trying to help them. To me, that's tough flying, terrible!

And if you've been flying a bunch of deer, the ice on your floats has frozen deer parts stuck in it, and it's red from deer blood. It's interesting; most pilots have never seen it.

The ropes are frozen; you can hold them out like a stick.

How about the splash rails—did you remove them? We did, because if you're breaking ice, it bends them.

I never remember taking them off. You got to remember, we had the PK 3500s the first years, and there was no splash rails. It was a piece that started at the nose and went all the way to the step. Those were the best floats to break ice with. Take 2870s, 2960s, that was tough because they were round-sided. If you got breaking a little bit of ice going ashore in a crosswind, you could dent the floats. The Edo 3430s and the PKs—listen, I've broke a lot of ice with my first 185, never had any dents 'cause of that. That was all new to me when I came to work. The first lost hunter I found, the second day I worked—George woke me up, I was living upstairs in the headquarters in Greenville. George said, "Hey, Jack's gone and we've got a lost hunter over at Squapan, you know where that is? Get ahold of Maurice Gordon." So I head over there. He told me where the guy had left from. Well geez, within twenty minutes I found him on the shore of Squapan Lake. He was there in a narrows and that was froze. And there he was by an old camp; he'd spent the night. So I knew he was all right. So I landed and I broke ice. I broke ice for a third of a mile. But it was not real thick. I remember breaking ice when the front of the floats would go up on, then break through, over and over. I remember Ray Porter—I'll never forget that—he'd have hunters right to the end and Ray would watch the ice. He had all these camps, Munsungun and stuff, and he'd tell them, "Tomorrow it's going to drop down to zero. We've got to get your people out!" And that was it, they had to move. You'd go by Shin Pond, 'course that would freeze, and they would keep a channel open right from their cove straight out. And if the wind was wrong, they'd taxi down and they had a place where they could turn around, in broken ice all the time. Right in the night, Ray or Clint would go out there in one of the airplanes and break it up so it wouldn't freeze so bad. And I can remember Jack landing on wheels on the ice and they were flying off floats in the channel. There was five inches of ice!

The ice demonstration I'll always remember was at our tent camp. We had a tent camp on Roaring Brook Pond over near Sebec. That tended to freeze early. We were there the first week in November. And we had radio communication, which made it a lot easier. Our hunters came out on Sunday. So we had one of those deals where we had bad weather on Saturday and a wicked cold front coming through that dropped the temperature way down. I was almost convinced by Sunday morning that the pond would be iced over. And sure enough, it was! So I talked to Max Folsom on the radio and I said, "Max, it froze last night!" Max replied, "Well, it can't be too thick; I'll come in in the Beaver." And I have never seen anything like it! When he landed in that skim ice, it flew, and there was ice sliding on top of the unbroken ice the length of the pond! There was no snow. And he taxied around—he wasn't having any trouble breaking it up. And then I think he made four trips out that day, and by the time he was done, there were just small pieces left.

We're talking about winter flying a little bit... Mac told me he had the PA-12 and he picked up a biologist in Shin Pond, and they were going to do deer yards up near Grand Lake Seboeis. He got off Shin Pond, and after five or six miles, there's [another] pond, Hot Pond, big enough to land on. And the oil cooler gelled right solid. There was oil coming out everywhere. So he set her down on the pond. They snowshoed back to Shin Pond, six miles in deep snow. That was just what they did.

Anybody who has never snowshoed in deep snow has no idea how much physical work it is. Usually, you try to switch lead men, one guy ahead for a while and then the next guy. But you are exhausted when you're done.

When I first came to work, John Leathers was a game warden. He was a pilot and he had the Yoke Pond country. So I had that yellow 185 and I picked him up, and he had his snowshoes and he asked me if I had some, which I did. We got up there and we see a flowage that was set up and he was sure something wasn't quite right. So we landed. There's a pond a little ways away called Pleasant Pond. So we set out snowshoeing. From the air, that didn't look too bad, about two miles. I was sore for a week! And the old fella—back then I was twenty-three and he was forty-five—it was all I could do to follow in his tracks!

I had the same thing happen with Bill Seely. Bill's snow machine broke down one time, and I flew him around in my Cub to pick up his traps. And he'd lead and I'd follow, and he was going so fast I couldn't keep up with him. When the day was done, I could hardly walk!

We were talking about some of the stuff we did with other agencies. Even today, we still do some flying for Maine Forestry. When I left Maine Forestry to go to Maine Fish and Game, I was familiar working fires, so I ended up flying some for them. I was checked out in the Huey and dropped water occasionally. Believe it or not, we did quite a lot of flying with the FAA. Every time there was a remote accident, they didn't have any way to get in there. I can remember an FAA guy: I knew him, he used to work for flight service in Millinocket, then he became an inspector. We had an accident up in that country, and they had to go in and investigate. So I flew him in there. It was not a good day and it became a worse day. It was snow squally, heavy, but you could fly. You could always see a couple or three miles. So we left the scene after he got everything he needed and went to Millinocket. So I'm going home because I can go down the interstate to Old Town, and I'm not sure what to say. Finally, I said, "I'm going home. What are you going to do?" He says, "Someone picking you up?" "No, I'm going to fly home. You want to go?" I said. Actually, it was probably legal, but it wasn't good. But I was comfortable, down the interstate, you know, you could always see a couple of miles in the snow. I'd rather fly in the snow than fog or rain anytime. You can always see straight down in the snow.

Another thing in the time period we're talking about, there were none of these cell towers everywhere. Now, this whole country is covered with towers. And you can't fly safely at low level.

They're everywhere! There are some…like there's one back of Shin Pond I discovered four years ago. I'm going into Shin Pond and it's just so-so, and I go by this valley between two hills and there was a tower.

These things literally go up in two weeks where there's never been anything before. So if you are a high-time pilot, and been around here a long time, all of a sudden there it is. I had it happen to me—you're going along and look off your nose and there is a tower.

You know who else we flew with, and we still do, but for a while we helped them out a lot—the Allagash Wilderness Waterway. They had no way to get around; they don't have fast boats, They use canoes most of the time. And we helped them out, tell them where people were. And they'd help us out.

Speaking of the Allagash Waterway, tell us a little bit about Otter Pond.

Allagash was so you couldn't land there, you couldn't use an ice auger, you couldn't use an electric shaver in there. So everybody would land on Otter Pond. I've seen ten airplanes on Otter.

Otter Pond is just outside the Waterway.

And Allagash is only open during February. So every year someone would bend something there, a landing gear, a wing tip, or a prop. They'd all land and fish in the same place in Allagash. No one is going to walk clear to the other end of the lake to fish. I remember people walking over another way. There's a pond to the east of it, up to where Johnson Stream comes in, and they'd land there and snowshoe across. It was about a mile and a half. But they wanted to fish that end of the lake. And then Johnson Pond in the spring and summer, you know, the air services went in there all the time. That wasn't an issue, but other pilots would get in trouble there.

Speaking of Johnson Pond, when the Allagash Waterway was created, people tried to figure a way they could still use airplanes to get to Allagash Lake without landing on it. But I can tell you, as someone who has flown into Johnson Pond many times, it is a very dangerous pond. When it's warm and the winds are swirling and you're using 180s and 185s, you can come out of there and really scare yourself.

There's been a couple of airplanes that have gone into the woods there. One is still there; it was a Champ way over in that southwest corner. I got a call one night around six thirty. The barracks called and said, "Bar Harbor Airlines on the way to Sherbrooke looked down, a guy that used to work at Folsom's, and said a boat is going around in a tight circle on Johnson Pond." Well, the Allagash fire tower man wasn't there or he would have seen it. They didn't think anyone was in the boat. So I jumped in the airplane and in forty minutes was there. There's the boat, but now it's on shore and the electric motor was running. And on the island, there was a campground there, a tent and one sleeping bag. So whoever was staying there must have been in the boat. The guy that was in the boat had [had] a heart attack and fell out. But because they were able to talk to the pilot, he gave them an idea of where the circle was. I think in an hour or an hour and a half, the divers had him. And you know, you were flying the divers out of there only one at a time!

Any other stuff we missed?

Winter flying then and now. You know, those first guys—I mean, that was tough flying. Andy Stinson flew twenty years, never had a heated hangar, never had a hangar, period. And now, the airplane I flew last year, I kept it right on the ice in front of my place and it has a plug-in heater. They wanted to give me a hangar up to Frenchville, but that's thirty-five or forty minutes. I kept my airplane outdoors at Old Town for the first eight or nine years because they did not want to pay for a hangar. I'd preheat it. I had one of those Red Dragons like you've got; still have it. That's the way it was. Now, it's a different story; they have nice equipment.

We went from the straight skis to the first Federals [skis with wheels] on the 180s. Those were 3200s. Then they went to 3600s—same ski but a little wider and some additional bracing on it. That is what they had when I come to work. Then we got the first set of the Fernandez, which, as far as performance went, was out of this world!

Describe that ski, the Fernandez.

The Fernandez was made out of fiberglass. It was designed by a guy in France, and when he decided he wanted to get into the market over here, he got hold of Folsom's. Dick said he wanted to send us a pair for free. Dick wanted to put them on Jack McPhee's airplane because he flew all day in the winter. So that's what we did. There were problems, like cycling the gear, because they were designed to take off on a nice runway and fly up into the Alps. They would only cycle [put the wheels up or down] in the air, whereas we had to go into the hangar and those weren't designed to do that. He did change them. When we got the next sets of them, they were for a Super Cub. And we thought they would be a scaled-down version. But when we took them out of the box, they were the exact same ski that went on a 185 except they had Cub gear. There's not a man on snowshoes that would stay up higher than that airplane did. Put on snowshoes and you sink down a foot next to the plane. They were out of this world! They were slow, the exactly same speed as floats.

Did slush bother it?

Not as much because you didn't get down into it. It had that little plastic bottom. We'd had plastic on the Federals; that worked pretty good, but this stuff worked really great. And then we went back to 185s and went to the Airglas [ski made in Alaska], I had the first set. And we bought them used. That's before they put the plastic bottoms on them. They were painted with that black paint. I found out that you could send those skis back to Anchorage and Airglas would put the plastic bottoms on them. It was peanuts to get it done. They were a completely different ski. It was fabulous!

Then last winter, the airplane I got had a new set of the Airglas twin tail, hydraulic, composite skis. That was very much like the Fluidynes with the twin tails. But it was also a copy of the Fernandez ski. And I flew that all last winter. It's quite a lot more money, double the price of the others. But for a whole winter, no issues, nothing. I flew them last winter about a hundred and eighty hours. They performed good; they were awesome. I'm sold on them! They've had those for quite a few years in Alaska and people from there had them, but no one from here had them. So I've had quite a few calls. They were twenty-some-odd thousand without a tail ski. The initial installation takes a while, plumbing going on, but it's a good ski. It's faster maybe. It's not faster than the regular

Airglas—you go one hundred fifty with that. These will go one hundred forty-five, no problem.

One difference is, the ski they copied, the aluminum Fluidyne 3000F, is heavy. Two of those weigh ninety-four pounds. My guess is that the carbon fiber could be half that.

I think it's sixty-something. The first time I saw those Fluidynes, you know who had them? Phil Walters from International Paper. And then, down at Pittsfield, the Cianchettes had an airplane with those skis on it. You know, speaking of the Cianchettes, they quit making 185s in 1985 and they got the last one. And one of the very last ones, Franklin Graham—got the big place up in Alaska—he bought it.

In the early days before your hangars, how did Mac and those guys do there switchovers?

Well, before they had Greenville, they would go to Waterville, land on the grass, and they'd put the floats on. They had a dolly. And the dolly wasn't like what we have now that's towed behind a pickup. They'd get the dolly out there, point the plane down the runway, and they'd let her go and use the air rudder. And when they lifted off, it was set up so when the weight got off it, the brakes would come on. The brakes didn't always work. They'd take out runway lights. And then when Greenville was in operation, everything was done there.

You mean once the hangar was built?

Yeah. I've stayed here on floats until the fifth of December or so, but I also have landed on the ice on the fifth of December. So that's how much it can vary. Same with Greenville. You know, you'd have to watch. George or Andy would watch and let us know it was going to freeze up and tell us to bring the airplanes in. So we'd take them out of the water. We'd try to keep one on wheels up to the airport so we'd always have something to fly. In the spring, occasionally we put the floatplane on the ice and took off on the ice. Most of the time, you waited till the ice was gone.

So when you get to today, you're on skis out here on Eagle Lake and you're getting ready to switch to floats. Where do they take the airplane?

They still go to Greenville. They use that facility. In the spring, they'll go up to the airport and use one of the hangars, put the floats on, and then take off on the dolly behind the pickup. And in the fall, they've gone back to going in, taking the floats off, putting the wheels on, and taking off on the five-hundred-foot driveway. 'Course these new guys, when you first told them about that, they didn't believe you. And of course, we done it, then Charlie done it. I can remember the first time I did it, holy geez! I can remember in the early '70s, Folsom would taxi over from their place, up the road, over, and back down. The sign that says Maine Fish and Game was set up so you could lift it up to get by. And I can remember three or four airplanes a day going out of there. I saw Dick Folsom taxi around with a Cessna Skyhawk, and he come down and he lined up near the road. It was December, there was a little powder snow, not enough to plow. He lined it up and set twenty degrees of flaps. There was a decent wind blowing. I watched him go, [and] the left wheel never turned. He went right out of there, kept it straight. I'm sure there was a little bit of drag there, but he only went a couple of hundred feet and was in the air. I landed a Cub there, you know, and I saw Charlie land a Cub there. Charlie landed a Cub there and the brakes weren't good. It was a south wind, which is what you want. What they did was to take a lot of the air out of the tires. He come right in and he hovered that Cub and didn't use half of it. I landed the surplus Cub there one morning. No wind at all. You just come in and put her on the end. You have to get it done!

'Course before Dick Folsom's house was built, they used to land there.

They started that after ours. They'd land there and take off there. They would come roaring over Howard's house. [The field] was a little longer than ours but had a bend in it. I remember Charlie and Ross Wheaton landed in there with a Cub and the seat broke.

The back of the seat broke off.

When he touched down, I think.

Well, the one I remember was different, maybe there were several incidents. The Cub had been rebuilt in the hangar and Charlie was going to take it off. So he went down to the end and turned around. And Charlie always had a habit of rearing back in the seat and twisting his neck before he took off. When he did that, the back of the seat broke off. So he didn't actually go up. There may have been another one. The other thing I remember, before Folsom's had a hangar at the airport, we used to taxi the planes from the airport the two miles down the East Road to the Lake to be put on floats.

I've done that.

And we would have a police escort. There would be lights in front and in back.

I remember landing outside of town, on Indian Hill. Bud Pelletier was the chief of police. It was Thanksgiving Day, 1976. I had a bad engine and they wanted me to bring it to Greenville. I landed in the road and taxied right down Main Street.

I'll tell you another story about Bud that I got a kick out of. I had my 180 in Greenville Junction, and I was taxiing it over to Folsom's to put it on floats. Since I didn't want to bother anyone, come daylight, I fired it up and headed off. When I got to Bartley's, which is halfway, I had flashing blue lights behind me, and I pulled over into Bartley's. Bud got out, and he says to me, "What are you doing?" And I told him I was taking the plane to Folsom's early so I wouldn't bother anyone. Bud said, "What do you think I'm here for? You never move that airplane again without a police escort." He got in front of me and put his light on for the second half of the way.

I've got a couple of other things here. I remember another accident down in Washington County, a guy in a P-11 down there. He was a trapper and had a guy with him and they never come home. So they didn't know just where they were; it was in January. And I got down there and I found the airplane. It had partially spun in and they were both dead. What happened, he had a beaver flowage set up right off a pond there. There was a little dead-water and they went around it. When the FAA investigated, they found an ice chisel had slid forward and locked the rudder pedals. I remember that well.

The thing about ice chisels is that they really should have a leather scabbard and be secured. Dick talks about seeing trapper planes land in Greenville with a chisel sticking out sideways through the fabric. They'll go right through fabric like nothing!

Remember what we always did, but they don't do it now, is to make a bracket to go on the rear float fitting. It was a tube. The end of the chisel went in there, and on the cabin step there was a bolt that went through the handle of it, secured with a safety pin. I wouldn't go anywhere without a chisel.

That was similar to the rig that George Later used to build for paddles.

I remember when he first built those. Matter of fact, Charlie gave me a set of those to put on this airplane.

Another accident…this was in Dick Varney's day. It would have been in the late '60s. They got a call about an airplane crash over by Lac Frontiere, over on the Quebec border near Saint Pamphile. The guy walked out and left the pilot there. The pilot was trapped in the airplane, a PA-12. The wreckage is still there. Well, early in the morning, Dick took off with my father, who was a Maine State Police officer then. They got there just as it was getting light, and just when they were near Presley Lake, the sky lit up. They got there and the light didn't last long. What happened, the guy was trapped in the airplane in a little bit of water and he had a flare. When he heard Dad's airplane, he lit the flare. Well geez, there was gas on the water! It didn't kill him, but his eyebrows and hair were singed. That plane hit the water. They were chasing a moose, and he looked behind and he cartwheeled. They knew what they were doing and told them. He owned the hotel in Lac Frontiere, which is on the Canadian side, but he was about three miles into the United States.

We had another—some of these happened before my time, but Mac told me. There was a guy trapping, he had a little bit of an air service in Portage. And in the winter he trapped. He had a J-3. He was in to Chase Lakes, over to the other side of Deboullie—there're three of them. He landed on Second Chase and was going to snowshoe down to Third Chase because he had beaver sets there. He was on skis. He never come home. So the third day, Mac gets over there and the guy has a fire going in an old camp on the end of the lake. He looks and he don't see the airplane. He looks again and there it was, rolled up in a ball. A big wind had come up while [the guy] was in the cabin and rolled that Cub right up. And it was laying on one side with one wing right straight. So Mac told me about that. Along about '85 or six, I go down there and someone was fishing the little pond. I landed and pull up to a ledge to tie the airplane, and right there was a whole bunch of dry-ki and I see something yellow in the water. It was a landing gear. I looked around, saw another piece of tubing, and then I saw a whole wing. The ribs were there but the spar had rotted.

Back in '74, I'd been asked to take one of our photographers out for a couple of days, and we were doing stories on beaver trapping in Northern Maine. 'Course, when you think about trapping, you think about Ray Porter. So I made contact with Ray and I told him what we wanted to do and we decided we'd meet at Churchill. Just off Churchill there's a stream, and of course, where he landed we couldn't land. We had to land, taxi up the steam, and then snowshoe to where Ray was. And we followed him around that day, two or three times and got pictures. So that night the photographer and I decided we were going to spend the night with Jack at his camp on Spider Lake. So we get there right before supper was on.

And around seven o'clock, there is a knock on the door. There was a guy that lived across the lake part of the winter. The guy said, "Hey, we've had a bad snowmobile accident across the lake. A guy's bleeding from the mouth and in tough shape." My airplane had been flown in the last few hours, so I took the wing covers off and taxied across. The sled had been going at a pretty good rate of speed and they hit a ledge. It threw the guys, and listen, they were both in tough shape. So they went back over and got Jack, and in the meantime, I took the rear seats out and we laid the guy that was bleeding bad on the floor. Then one of his friends got in with me. So we taxied away from shore. It's snowing, not hard, but enough that it wasn't great. About then, here comes Jack across the lake. He's going to pick up the second guy. I turned—there was a little bit of wind out of the southwest—and lined up going to the southwest where there was two hills off to each side. I took off, get up to a couple of hundred feet, and brought the power back. I turned the landing light on to see how hard it was snowing. When I did, you couldn't see out the front because of condensation, and when I turned to the side, I was looking at the side of a hill where there was an active cutting operation. When I took off, I had veered to the left some. If I had kept going straight, we probably would have gone into the woods there. It was a pretty steep hill. I've looked at it hundreds of time since then. There's a good example: I should have took off towards Churchill Lake—flat as a pancake over there.

So I got up and I started for Presque Isle. I called the State Police barracks, told them what we had, to get the ambulances there. That's before GPS and Loran. So I go east and there is nothing in that country. That area in the night is pretty black. We went by one lumber camp with lights. And about the time we got to Portage and Ashland, we started to see the lights of those towns. Within seven or eight miles of Presque Isle, I could see the ambulance lights at the airport. And that was Northeast Airlines at the time, and they held them up till I landed. In the meantime, Jack loads the other guy and takes another guy with him. Jack takes off, must have turned a little, and in the landing light he sees an island. He put it up, and the next day he could see where the right wing brushed the snow. He showed up maybe fifteen minutes after I did. And one of those guys died. Both were tough; I don't know if it was the guy I had or the one Jack had. My guy had a lot of jaw damage and his mouth was torn up. He kept gagging, and I told the other guy to reach in with his fingers to keep his air [clear]. That was in February of 1974. Now, we just happened to be there. We could have been in Greenville that night. And Jack might not have been there. If nobody was there, both those guys would have died. The nearest plowed road was like fourteen miles away. I will always remember that. I've looked the hill over many times since then.

You know, I'm the same way. I can still remember one close call coming out of Johnson Pond. For years, if I was close, I'd look at it. Do you feel the same way?

I do the same thing. And often, as I am flying along, I'll remember that I picked up a guy there because, like I told you, I have found over two hundred lost or drowned people. 'Course, we used to have a lot of them! Now, it's just not like that anymore.

Another search I had, one of the last years I worked, we had a guy get lost out of Scraggly Lake over by Millimagassett. And the guy had a good start on us. So we got over there and the lakes were frozen; it was the last week of hunting. So we're looking, and I happen to look by Ed Stone's camp—there's a set of tracks around the camp where he had walked along the shore. That's got to be our guy; there's no one else around there. I see where he headed east from there, he left the lake. I never saw tracks again and I kept looking. I didn't know if he was two hours ahead of us or twenty hours ahead of us. Well, right before dark I found him. Finally, he built a fire, and he was down in a gravel pit off a road on Grand Lake Seboeis. That is a long, long ways away. He done a great circle route, I call it, and built a fire. I had the fire in five or six minutes. We figured he walked fourteen miles!

Were you on the search for the guy that ended up on Wilson Pond within sight of the Greenville airport?

There's quite a story to that. I never worked on that search. Tim Tarr did, Jim Welch did. That day we had fourteen different people lost who had spent the night in the woods. I found two. I found one on Nollesemic Lake, just outside Millinocket, and I found the other guy over near Haynesville. You know, when you got fourteen guys, you can't go to all of them. Well, the guy on Wilson Pond didn't walk out. He was the one who said he could sit in the camp and see the airport beacon. Well, we had stuff like that over the years. If I had been in a camp and could see the beacon, after the second day, they would have come to see what was burning. I would have started a fire. I met the guy a couple of years later. But I didn't work it. Back then, we had lost hunters every morning.

I want to mention the pilots since day one. I think this is all of them. Now, you got to remember that over the years we had wardens who were pilots who flew to help us out—Eben Perry, Gray Morrison, Maurice Gordon. They usually flew the Cub. Gray used to fly the 180 or 185. He helped us fish stocking a little. So there was four or five wardens over the years that helped out, a guy named Earl Kelley, he was one. The first pilot was Bill Turgeon. Turgeon flew in World War I, and when my father got into the flying business, he'd go down to Marden Airways in Waterville. And that's where he met Howard, and that's where Maine Fish and Game was having their work done. And he roomed one whole week with Bill Turgeon. Then George Later got hired, then Malcolm "Mac" Maheu. The story on Mac—when they hired him, he was out of the military. He used to pilot airplanes for that deal where they'd take them over to Russia. Mac had been around! He learned to fly at the same school where Lindbergh learned, out in the Dakotas or somewhere. Mac was jacking up a camp in Rockwood, he had sporting camps there. And George and Bill Turgeon, they went up to see him because they wanted to hire a pilot and they knew what he was! So they get up there and Mac is under the camp. And he looks out through a crack and sees two game wardens and didn't know what they were after. And that's where they hired him.

Sleepy Atkins—real name was Wilfred—he got hired about the same time, maybe a little before Mac. They were all good, good friends. George Townsend was one of the pilots we had who flew out of Rangeley. We had a base there. He flew until he had his accident. Bill Snow, he worked for us for five or six years, then he went to Federal Fish and Wildlife. Andy Stinson came to work to replace Bill Snow. Dick Varney came to work and he took George Townsend's place, except Dick was sent up here. After Dick Varney, Jack McPhee.

Jack got hired in '67 when I was learning to fly. I had soloed. And I remember a big thing. I would fly over to Eagle in my dad's PA-11 and Jack would be here. The first time I met Jack to talk to, he had a brand-new Super Cub. He said, "I'm going to Shield's Branch; you want to go?" Well, I didn't know where that was, but I was going! He said, "I got to go drop some gas off." He had four of those five-gallon military-type gas cans, the square ones. He put them right in the baggage compartment. So we get ready to push off and he says, "Get right up front, I'll fly from the back." So he showed me about the flaps. I had never been in an airplane that had flaps or an electric starter. He put me right in the front seat and we took off and went up to Shield's Branch up on the Big Black. He put the fuel behind an old camp there, and that was my first meeting with Jack. He let me fly the Cub. So that was '67, my junior year in high school. He was part of the reason I looked into going to an army flight school. He talked to me about it. Jack had been a warden before he was a warden pilot.

So after Jack, I came along and Dana Toothaker came along. Then in the late '70s, Jim Welch, Roland Tarr, who was from Blue Hill, been a warden a long time. That's about when we went to a bunch of Super Cubs. So for a while we all had Super Cubs except Dana, who kept a 185. Then Alan Ryder took my place when I retired in '91. Then we had Roger Wolverton, Jason Bouchard, and then Durwood Humphrey. I knew Durwood since he was in high school. He worked at Old Town airport. He flew out of here for quite a while. Then they hired Charlie Later, then Dan DuFault, then Daryl Gordon. Then four or five years ago, they hired Jeff Beach and Jeff Spencer. Jeff Spencer is up here in Eagle Lake now. And last year we hired Chris Hilton. He was a Maine Marine Patrol officer and commercial pilot. He is Rodney Folsom's son-in-law. I don't think I forgot anybody. Those are the pilots since day one.

So in that list, when did the pilots first have to go to warden school?

Everybody went except for Bill Turgeon at some point or other. I remember Andy going. 'Course, back then it was six or eight weeks. I never went for thirteen years. What would happen is, every time they had a school, it was usually in the spring. And I'd be scheduled to go to it. And back then, we used to have a big problem with dogs chasing deer. That's something we didn't talk about. It was a big to-do in the spring of the year, down state especially, where there were a lot of deer. It wasn't a dog problem we had, it was a people problem. They didn't take care of their dogs! So every year when I'd be scheduled to go to school, they'd tell me at the last minute not to go. 'Course, we were using the helicopters, so I never went for the first thirteen years.

So what did you do when you came across dogs chasing deer? What was the solution?

Well, technically, if they're chasing deer you shoot them, simple as that. We'd shoot them out of the helicopter. We didn't do it a lot;

you had to be careful. Now today, I don't know if that would fly. A lot of the time, we'd follow the dog home. Many times it was a puppy, wouldn't have even hurt a deer. But you know, if you run a deer long enough on crust it will do damage. Listen, I remember down to Old Town and Orono, about '80, we drug out nineteen deer right in one area. So you get four or five big dogs out there that know what they're doing and it's trouble. That's slowed down and I like to think we educated the people.

Out of all those pilots, there are four of us who never worked as a game warden. I went to work day one as a pilot; Bill Turgeon did; Mac Maheu never worked as a district warden; and Andy Stinson never worked as a district warden. But generally, they would hire from within if they had people who were qualified. It's getting harder and harder to get pilots. I can remember when a warden pilot job was open in the early '70s, there would be a lot of interested people. They'd have to cut it down to three of the best. Now they have a hard time to find three that qualify. In the old days, there were a lot of people that flew floats and skis and worked for Folsom's. You just don't see that anymore.

Starting at the beginning, were the flying wardens armed?

I would guess so. I look at the old pictures and see guys in the '50s with guns. Listen, I don't think I ever flew with a gun. I didn't wear it a lot. Probably that wasn't right some of the time. It was just cumbersome in the airplane, getting in and out in a hurry having a gun belt on. I don't think Andy wore a gun much. We just flew. We did our thing and knew what to do. But you had to go to school, and as it got along, when I finally went after thirteen years, it was an issue, my liability. Let's say you got into a problem with an armed standoff and you didn't have any training, the Service had to look out for themselves. Now, nobody goes to work without the training. I remember a game warden years ago, worked four years before going to school.

Have the wardens ever had upset training under water? You know, where they have a cockpit that rotates in the water.

No, we've never done that. We've talked about it. I have a good friend—never met the guy; he's originally from New Sweden, Maine. He's a pilot for the Alaska State Troopers. The Alaska Troopers, they have the green shirts and the blue shirts. The green shirts mean the wildlife trooper. This guy is a pilot for them, has been for over twenty years. And we correspond quite a lot. 'Course, they got a lot of pilots. There're doing some training. A lot of their pilots are wardens who have an airplane. [Each guy covers] eight to twelve villages. They're not actually a trooper pilot; the plane is the way they get from village to village. They have a few pilots that all they do is fly. They have a hell of an air force. They still have a fleet of Super Cubs. They're Super Cub people. They have their own fuselage that's wider now. They rebuild their Cubs. They say when you're out chasing bear hunters, you need a Super Cub to get anywhere near.

They got other airplanes. They got 180s, 185s, and 206s. They have a Caravan, maybe two. They use those to move prisoners from the villages to bigger towns. And Scot Quist, he has been a pilot for them for a long time. He's also helicopter rated. They have quite a few of those small R-44 helicopters. And they have a couple of those Aerospatiale turbines. And they had a bad accident four or five years ago. A couple of guys died. So they have so much going on, you have to do training.

We're doing different training. Our guys go south and do instrument training so they can keep an airplane right side up if they get in a bind. But as far as you're talking about, they had a program somewhere where they put you in the water in an airplane wreck. Then they turn it upside down and you have to get out. I've never been, but I've talked to people, and they say you can't imagine how dark it is in a Cessna upside down in the lake. And I can think of accidents over the years where people... Before my time, there was a 180 with two guys on Harrington Lake. They took off with glassy water, they settled, and then they cartwheeled. By the time they got out there with a boat, both those guys had drowned. They got tangled up in their own rope. You know, I taught float flying for a long time, and you don't get in an airplane with loose rope. I don't care if it is a twelve-foot piece, I coil it. And also, I always have a piece by the door that I can grab when I get out. That's automatic! And in the winter, my axe is right behind my seat. I have a piece of Velcro where I keep it. Another thing I do in the winter. I have been through the ice twice. I have "the picks of life"—you must be familiar with those. I fly with a vest in the winter, and in it are the picks, some fire-starter material, some waterproof matches, and a little lighter. You know, you might get out and that's all you got. Like Mac, if he hadn't had that camp there, he probably would not have survived. Quite often in Alaska, someone will go through the ice, and when the airplane settled to the wings, they couldn't open the doors. So when you think you're starting to go, unlatch it. Jeff did that out there this winter—he unlatched the doors and he got out. 'Course he couldn't jump straight out because the wings are there, so he jumped at an angle and got up on the ice. It held him, an inch and a half of ice. And he felt with his foot and climbed right up on the ice.

The people that run that upset training, they claim that a lot of people have a tendency, once they get free of the airplane, to swim for the bottom because they are totally disoriented, and that's one of the things they teach—to look for the light.

I forgot who I talked to that said you can't imagine how black that is!

The other thing that is a problem is when you're upside down and the seat belts are still fastened. It's a different load on the belts and they are very difficult to get off.

You know, talking of seat belts, that brings up stories. Mac had gone and picked up a guy that had been involved in an airplane accident—he'd hurt his shoulder. Anyway, Mac picked him up, flew him back to Greenville. And the Maine Fish and Game hangar then was where Maine Forestry is now. Mac came in with the wind howling, went to turn into the tie-down, and the wind picked up and flipped the plane right over. The guy said, "Geez, I survived two accidents!" It was probably blowing thirty, and in a Cruiser or a Cub, you have to be careful.

You think that's about it?

G: Well, you know, I guess.

Well that's a lot! I want to say you've done an absolutely great job! Great memory! Thanks, Gary.

Mac adds oil to a 125 horsepower Piper PA-12.

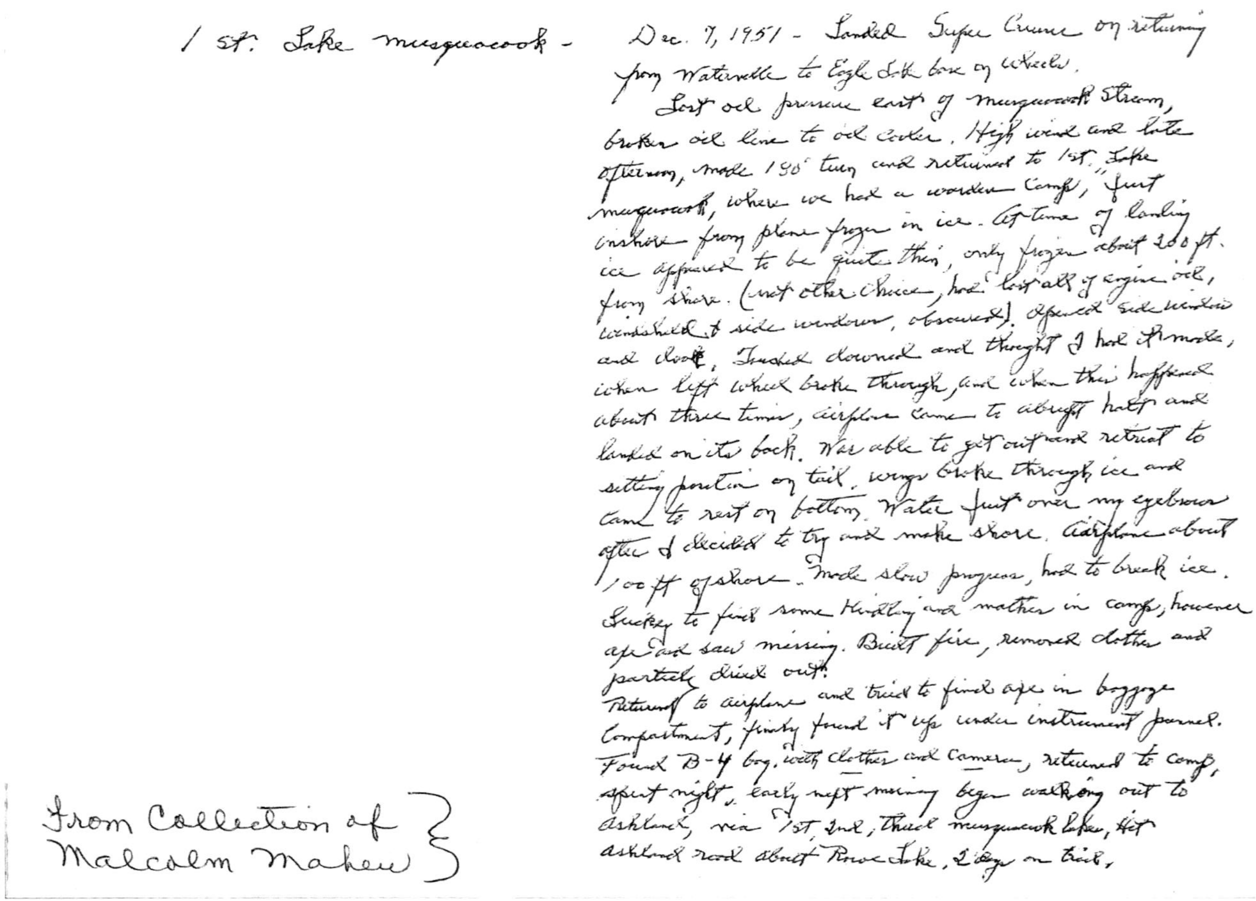
1st. Lake Musquacook - Dec. 7, 1951 - Landed Super Cruiser on returning from Waterville to Eagle Lake base on wheels.

Lost oil pressure east of Musquacook Stream, broken oil line to oil cooler. High wind and late afternoon, made 180° turn and returned to 1st. Lake Musquacook, where we had a warden camp, just inshore from plane frozen in ice. At time of landing ice appeared to be quite thin, only frozen about 200 ft. from shore. (not other choice, had lost all of engine oil, windshield & side windows obscured). Opened side window and door, landed downwind and thought I had it made, when left wheel broke through, and when this happened about three times, airplane came to abrupt halt and landed on its back. Was able to get out and retreat to sitting position on tail, wings broke through ice and came to rest on bottom. Water just over my eyebrows after I decided to try and make shore. Airplane about 100 ft of shore. Made slow progress, had to break ice. Lucky to find some kindling and matches in camp, however axe and saw missing. Built fire, removed clothes and partially dried out.

Returned to airplane and tried to find axe in baggage compartment, finally found it up under instrument panel. Found B-4 bag, with clothes and camera, returned to camp, spent night, early next morning began walking out to Ashland, via 1st, 2nd, Third Musquacook lakes, hit Ashland road about Rowe Lake, 2 days on trail,

From Collection of Malcolm Maher

Mac's story in his own hand.

Mac's Piper PA-12 on First Musquacook Pond.

A hermit who lived near the Baker Branch on the Saint John River.

Maine Fish and Game base on Moosehead Lake. (note the short runway behind the hangar)

The Maine Warden Service Stinson Gullwing after maintenance.

First snowfall covers a Piper PA-12 at the Eagle Lake base.

Mac's Piper PA-12 after landing on the rough Daaquam air strip.

A Maine Warden Service Aeronca Sedan ready to haul a cookstove to the Clear Lake camp.

An early "jitterbug" on Umsaskis Lake.

Three men were lost when this Piper PA-12 crashed on Sabeo Lake.

The Poland Pond fire.

Al Nugent, founder of Nugent's camps on Chamberlain Lake, helps out on the Poland Pond fire.

Hayne's trapper camp 1950.

Trapper Haynes.

Andy Stinson, Maine Warden Pilot, at Jean Duprey's cabin on Allagash Lake.

George Later at the Elephant Mountain B-52 crash site shortly after 1/23/1963.

Maine Warden Pilot, "Sleepy" Atkins.

Mac Maheu, center, guards Eisenhower's limousine during his 1955 fishing trip to Maine.

Eisenhower arrives in Maine.

Maine Warden Pilot, "Sleepy" Atkins, examines the Scribner crash on Desolation Pond.

Hoisting a Piper Super Cub on Moosehead Lake.

The first Maine Warden Pilot base in Greenville, where Maine Forestry is located today.

A Maine Warden Service Piper PA-12 which flipped on Brassua Lake.

Early fish stocking using buckets and a cabin fish tank.

A Maine Warden Service Piper PA-12 sitting on the "runway take off dolly."

Crash at Churchill Dam.

Broken axles or landing gear were common during ski flying in rough conditions.

A floatplane, tied to the dock, was flipped by an intense squall on Moosehead Lake.

Mac's Piper PA-12 at the Simond's fox farm on the Saint John River.

Umsaskis Lake warden camp 1950.

Preheating Mac's Cessna 180 at the Eagle Lake base.

A Maine Warden Service Piper Super Cub with Fernandez skis.

Budworm DC-4 spray plane after a forced landing on Churchill Lake.

Fish stocking tanks designed and fabricated by warden mechanic Howard Lambertson.

Maine Warden Service divers searching during the winter.

Gary Dumond taxiing through Greenville after landing on Route 15.

Maine Warden Pilots survey ice thickness.

Maine Warden Service mechanic Howard Lambertson.